**ALLVINE, Fred C. and James M. Patterson. Highway robbery; an
analysis of the gasoline crisis. Indiana, 1974. 261p il tab 74-
1598. 10.00. ISBN 0-253-13750-0. C.I.P.**

Updates the authors' previous study, *Competition, Ltd.; the marketing of gasoline* (CHOICE, Feb. 1973). The emphasis is on the governmental policy decisions under price and import controls (pre-Arab oil
crisis) up to mid-1973. Highly readable, this volume traces the mis-
takes of governmental regulation and the reaction of oil companies.
The economic analysis, weak in spots (some conclusions are not supported by the best evidence or arguments), is readily understandable
by any freshman who knows that price and quantity demanded are in-
versely related. As descriptive history, a very good volume. Allvine
assisted Senator Jackson on the Senate Permanent Subcommittee on
Investigations while it investigated petroleum shortages, hence the
factual basis for this book is strong.

Highway Robbery

Highway Robbery

AN ANALYSIS OF THE

Gasoline Crisis

Fred C. Allvine and James M. Patterson

INDIANA UNIVERSITY PRESS   BLOOMINGTON AND LONDON

SECOND PRINTING 1975

Published in Canada by Fitzhenry & Whiteside Limited,
Don Mills, Ontario
Manufactured in the United States of America

Library of Congress Cataloging in Publication Data
Allvine, Fred C.
    Highway Robbery

    Includes bibliographical references.
    1. Petroleum industry and trade—United States.
    2. Petroleum products—Prices—United States.
    I. Patterson, James Milton, joint author.   II. Title
    HD9564.A79   1974   338.4'3665'538270973   74–1598
    ISBN 0–253–13750–0

Contents

To
Ron J. Peterson
Intertype competitor, industry statesman, and friend

List of Tables

List of Figures

Preface

In 1969 the authors of this book began a study exploring why the gasoline industry was one of the last holdouts in the adoption of more efficient mass merchandising techniques that had rapidly penetrated other industries. This research led to the publishing of *Competition Ltd.: The Marketing of Gasoline* in 1972. A primary finding of that book was that vertical integration was largely responsible for the archaic and wasteful system of marketing gasoline which had failed to respond to the diverse and changing preferences of the motoring public.

A key point made in *Competition Ltd.* was the importance of "intertype competition"—competition between the large integrated petroleum companies with their nonprice methods of marketing gasoline and the independent gasoline marketers, who sold gasoline at lower discount prices. It was emphasized that, if this intertype competition was protected from the unfair, functionally diversified financial power of the integrated giants, there would be a gradual revolution in gasoline marketing methods that would benefit the public. Intertype competition was successfully challenging existing methods, regulating excessive retail margins, and responding to new preferences.

Vertical divestment was proposed as a way of enhancing market competition. However, given the political reality of that period, it was considered an unrealistic remedy. Practical emphasis was placed on declaring certain competitive practices used by the majors to destroy intertype competition as "unfair trade practices," in the sense that this phrase is used in Section 5 of the Federal Trade Commission Act.

In a sense we underestimated the very great impact that intertype competition was having on the gasoline industry. At the very time *Competition Ltd.* was being published, the major marketing style was being challenged at its very core. The way the majors were going to respond was already taking shape. The engineering of a product shortage had already begun. The destruction of intertype competition was in the mill.

By late 1972 and early 1973, the independents were being decimated. Intertype competition had been effectively stopped. Forces for retail change had been neutralized. Price wars had ceased. Retail margin had started an unprecedented upward climb.

This was a whole new ball game. Intertype competition was no longer being "limited"—it was being neutralized and destroyed. The polite mayhem of the 1960's was being displaced by a concerted effort to drive the innovative intertype competitors out of the business altogether.

In June 1973 Professor Allvine was asked to work with Senator Henry Jackson on the Senate Permanent Subcommittee on Investigations which was examining the causes of the recent petroleum product shortages. This subcommittee's work in uncovering state and federal information on the industry, which was not generally available to academic analysts, provided much of the factual basis for this book.

The authors acknowledge and thank the Council of Independent Gasoline Marketers for the research grant made to Georgia Institute of Technology in support of this study. We are also grateful to the many firms—independent and semi-major—who were willing to discuss the details of the gasoline shortage with us. Many thanks also go to Miss Elaine Preston who typed several drafts of the manuscript and to John Connelly, Steve Lacy, J. D. Alexander, and Michael Larimer who were research assistants.

We especially wish to acknowledge the influence of Ron Peterson of Martin Oil Service Company, to whom this book is dedicated, and Senator Henry M. Jackson, whose leadership of the Senate Permanent Subcommittee on Investigations gave us the facts to support our suspicions.

Finally, the authors accept total responsibility for the views expressed in this book. The preparation of the manuscript was kept highly confidential and was neither read nor reviewed by industry prior to its publication.

Highway Robbery

1

Where Have All
the Green Stamps Gone?

The summer of 1973 marked a major turning point in the history of the oil industry and in the marketing of gasoline. Johnny Cash had been telling us all spring about the virtues of Amoco gasoline and its dealers. Suddenly he started telling us to drive more slowly and less often. Shell's advertisement of a car driving through the paper barrier was at first replaced by Shell tires breaking the paper barrier and later by Richard Basehart offering sober advice on how to save gasoline by getting tune-ups and checking tire inflation. Fortunately, there were no major space shots or political conventions, so Gulf was spared the problem of figuring how to *not* sell gasoline every twelve minutes for hours on end. Needless to say, consternation reigned on Madison Avenue and in petroleum product sales offices.

Several oil companies on key turnpikes began to limit sales to ten gallons per car, and friendly neighborhood dealers began going home to watch television in the evenings and drive their own families out to the lake on Sundays. Kansas City and Denver had major Sunday afternoon crises with two hour lines at open stations. Even the AAA began to issue gasoline reports along with its traditional travel reports.

No doubt, the high point in public disaffection was reached when an enraged motorist fatally shot a service station attendant in Oakland for refusing to pump more than ten gallons of gasoline into his car.

By June, Texaco, American, Gulf, Sun, Shell, Exxon, Phillips,

Mobil, Union, Arco, and Citgo were rationing gasoline to their dealers
—some to as little as 80 percent of last year's sales.

Later in the summer the gasoline shortage was joined by a
threatened heating oil shortage. "Dial-down" became a companion
slogan, as the nation braced for the expected winter surge in the
demand for distillates—along with gasoline, the other major product
refined from a barrel of crude oil. The expectations were that we
would have to import 500,000 to 600,000 barrels per day of the middle
distillates (No.2 heating oil and diesel fuel) if we were going to be
able to make it through the winter—and that amount only if the
winter were mild.

Energy czars succeeded each other in rapid succession as the
President periodically appeared on network television to reassure the
man in the street that we would be able to manage through the winter
if only we would all do our part. Driving to grandmother's house was
now almost an unpatriotic act.

On October 17, 1973, the whole prospect changed. The threatened
gasoline and heating oil shortage now became an oil crisis as the Arab
members of the Organization of Petroleum Exporting Countries
(OPEC) imposed an outright embargo on those nations (including
the United States) that had shown insufficient enthusiasm for the
Arab cause in the October War with the Israelis. This embargo was
compounded by the simultaneous programmed cut-back of crude oil
production which further reduced the free world's supply of crude
and thereby reduced the chances of third-party transactions by-
passing the embargo.

Consternation reigned. The Northeast was especially apprehen-
sive, given their dependence on oil heat and residual oil for electric
power generation. Front page stories about heating oil and gasoline
shortages and rumored gas rationing were widespread. "Energy experts" of all persuasions began to come out of the woodwork. For a
while they were replacing diet experts and sex counselors as television talk show guests. No show was complete without one. Every
progressive Rotary Club program chairman needed to devote at least
one Wednesday to the "energy crisis" and, if he was really progressive, he would schedule two programs so he could give "both sides."

In an industry which has long been known for its marketing overkill, such reversals made a major impression on the public. Abstract
arguments about wasteful marketing practices and ill-founded public

policies which were needlessly costing the American public billions of dollars cut little ice and aroused little or no public indignation. Under pressure from the American Petroleum Institute (API), President Nixon could even override his own blue-ribbon Cabinet Committee's recommendations to terminate the mandatory oil import quota program which had long since ceased to function in the public interest—if it ever had—without the slightest concern for adverse political repercussions. No one cared; there were more important things to think about. What's so bad about another industry feeding at the public trough? Everyone does it, one way or another. And salad bowls, trading stamps, and clean restrooms were nice to have and not harmful to anyone. Besides, the difference between 36.9¢ and 38.9¢ is only a couple of pennies.

Then came the shortage. This was real. It interrupted vacations and threatened life styles. The papers now began to talk about gasoline at $1.00 a gallon and a ten cent tax increase. City after city began having trouble finding suppliers who were willing to sell them gasoline for police cars and garbage trucks. School systems could not get long-term suppliers for their buses. Farmers and truckers began to have trouble getting short-term gasoline and diesel fuel. Florida and other vacation areas found their tourist business dropping off. Winnebago cancelled plans to build a West Coast plant and drastically cut its production in Forrest City, Iowa. Even Amtrak began to get customers, and decaying depots came alive once again.

Independent private brand marketers found their traditional sources of unbranded gasoline cut off. Some independents were operating at 60 percent of their previous volume or less. Many had to close. Most had to raise their prices to try to make up for the lost volume. All faced a bleak prospect as supply contracts ran out and as unbranded supply became even tighter.

Hundreds of small, independently owned, branded jobbers who had represented major brands for years found themselves cut off from their traditional suppliers as Gulf, Phillips, Arco, and others began to consolidate their marketing territories. In normal times such market withdrawals would have meant little, but in time of shortage, it meant these small businesses—many of which had expanded at the insistence of their former suppliers—would be unable to find new suppliers and, hence, would be forced out of business.

The great gasoline shortage of 1973–74—which was only a threatened shortage until the Arab embargo—did what no other argument or event had been able to do. It riveted public opinion into an instrument of reform. Six state Attorneys General marched on Washington, and two subsequently filed major antitrust suits seeking divestment and treble damages from the major oil companies. Others are following suit. After sitting on its investigations of the petroleum industry for several years, the Federal Trade Commission finally brought a major antitrust action against the "big eight" oil companies.

Even the sacred Mandatory Oil Import Control Program was scrapped and replaced by an open-ended license fee program. The Trans-Alaska Pipeline, which had been stalled by the environmentalists, was revived by a close Senate vote.

Dozens of law suits were filed by cancelled dealers and jobbers asking for restraining orders against supply termination and antitrust damages. Lawyers on all sides will be busily engaged for years.

On the West Coast hints of conspiracies caused a grand jury to be convened and oil company executives were kept busy answering subpoenas and explaining to reporters how they were as innocent as Caesar's wife should be. Xerox machines in public relations offices all over the country were running overtime, grinding out press releases about the virtues of private enterprise.

More important than any other factor, the Congress began to stir. For the first time in several decades, Congress seemed to be in a mood—shielded from the oil lobby and the White House by Watergate, and prodded by an angry and disenchanted public—to reform the structure of the industry so we would not run out of gas again. Divestiture bills to break up the big companies which had been common in the heyday of the "New Deal" were revived in both houses. Tax reforms and windfall profit taxes were widely discussed. Powerful committee chairmen began to take an interest. Hearings and staff investigations were begun. Interest groups began to descend on the Capitol from all directions. The fight was on.

Environmentalists properly sensed that, under the guise of an "energy crisis," hard won gains of earlier years could be lost. The swiftness of the Senate action on the Trans-Alaska Pipeline emphasized the vulnerability of clean air standards.

Transportation, agriculture, construction, and other industries, sensing fuel shortages and consequent disruptions, all showed up with special pleadings for "preferences."

State and local agencies and other public bodies also showed up with documented horror stories about summary cancellations. They volunteered elaborate testimonials on their indispensable service to society.

Independent gasoline marketers, refiners, and terminal operators, oil jobbers and organized gasoline retailers all cranked their Washington operations into full swing.

In short, the way gasoline and other petroleum fuels had been produced and distributed was being fundamentally challenged. That it could be handled differently in the future appeared to be possible. Whether the gasoline shortage of 1973–74 was contrived or merely the result of poor planning is irrelevant. The public is in no mood to let it happen again. The fact that a small group of large, vertically integrated,* international oil companies could decide not to build additional refining capacity in the face of growing demand and even profit from this decision does not sit too well with the man in the street. He wants things changed and he may get his way.

On the other hand, the way the industry should be changed is not altogether clear. The issue is further clouded by the smoke screen raised by the long list of special pleaders. The energy industry, of which the petroleum and gasoline industries are merely parts, is exceedingly complex. Short-term solutions can lead to long-run disasters. Solutions to problems in one area can result in more serious problems in others. Issues of national security and economic well-being are at stake. The quality of our environment and the character of our life style are involved.

Why Is There a Gasoline and Distillate Shortage?

For many, the answer to this question is that it was caused by the Arab embargo. The fact is that the embargo merely exacerbated the shortage; it did not cause it. We had already experienced shortages before the embargo was imposed. We were going to have trouble getting through the winter of 1973 and the summer of 1974 even without the embargo. Obviously, the oil companies would like to use the embargo as a scapegoat to divert public attention from the need to

---

* Vertical integration, in the static sense, characterizes those situations where a single firm is involved to varying degrees in raw material production, manufacturing, and distribution or various combinations of these functional activities.

explore the basic structure and operating performance of the petro-leum industry.

Just as the Arab embargo did not cause the gasoline and distillate shortage, neither will its removal eliminate it. Actually, the reason for the short-fall of gasoline and distillate inventories is quite simple: U.S. refining capacity has not grown apace with the rapidly rising demand for refined products. But the reasons why refining capacity has not kept pace with demand are more complicated. A multitude of theories exists. Some see the slower growth of refining capacity rela-tive to demand as part of a covert conspiracy—or at least as part of a conscious pattern of parallel behavior by the major oil companies to get rid of excess refining capacity in order to eliminate the competi-tion of independent marketers, pressure changes in public policy, and ultimately to raise prices in order to generate funds necessary for the massive capital demands faced by the industry during the next ten to fifteen years.

The counter theory advanced by the oil companies is that the reasons lie in the confused and uncertain public policies which made long-term extensive investment in new refining capacity unattractive. Also, environmentalist opposition to refinery siting, uncertainty over the use of lead additives to raise octane ratings, uncertainty over what required octane ratings will be, oil import policy, how the Alaskan oil would be delivered, low returns on refining, and other factors are cited as reasons for going slow on refinery construction. What kind of refinery to build and where to build it depends upon the answers to these key uncertainties. Thus prudence, not conspiracy or imitation, is the reason for holding back.

Still others argue that the shortage was a planning mistake. De-mand, it is argued, grew faster than expected and a series of simple miscalculations, coupled with long lead times and some bad breaks with the weather, turned what would otherwise have been a good management decision—to keep capacity expansion in tune with de-mand growth—into a management mistake, producing spot shortages in certain areas and in certain channels.

No doubt each of these theories is valid to some degree. Yet per-haps the most important explanations lie in the structure of the in-dustry. If so, the whole nature of the problem changes. No longer are we concerned with the fact that a small number of large, vertically integrated oil companies failed to build adequate productive capacity in anticipation of rapidly rising demand. The real issue now becomes one of *power*. That this small group of men *could*, whether by design, prudence, or error, hold back on the expansion of productive capacity in this vital area, without anyone else able to do anything about it, is itself a social peril.

Recent history shows that only crude-rich and vertically integrated oil companies dare enter the refining business. Nonintegrated refiners would be foolish to do so in the face of uncertain public policies and the economic power of integrated competition. With this knowledge one can see why oil companies, lacking *both extant and potential competition have failed to build adequate refining capacity.* Moreover, if this theory is correct, the solution to the problem lies in changing the structure of the industry in order to promote competition, rather than in the encouragement of better planning and more responsible behavior on the part of that small group of businessmen who now govern the flow of oil.

The actual sequence of events leading up to the shortage of refining capacity in 1973 makes it hard to understand why it occurred unless the majors were in fact trying to force a showdown. It was predictably only a matter of time before demand outran supply. If not 1973, it would have been 1974 or 1975.

In fact, the shortage came as no surprise to the majors. It had been predicted in numerous speeches by major executives over the last two or three years.

In early 1971, the authoritative *Oil & Gas Journal* stated:

> During the Seventies, domestic refiners will be forced to take on a heavier share of burgeoning demand for energy. . . .
> If refiners generally do not cut through their many problems and close some new construction contracts this summer, they will face difficulties they haven't dreamed of. Since it takes two or three years to complete any modern refinery, a crunch on domestic petroleum products appears almost inevitable by the mid 1970's.
> —(*Oil & Gas Journal*, March 22, 1971)

The demand for gasoline has been rising steadily over the past decade, from 66.3 billion gallons in 1962 to 102.6 billion gallons in 1972. This is a 54.9 percent increase. True, the growth has accelerated over the last three or four years, but even this acceleration was easily explained and behaved according to easily predictable rules. It is due to three basic forces—*more cars* driving *greater distances* each year while getting *fewer miles per gallon*. Furthermore, none of these forces showed signs of quick reversal. More cars are being sold than ever before. The new cars, because of increased weight, air conditioning, emission controls, and other accessories, are getting less mileage per gallon. Finally, economic expansion with its attendant growth in affluence and leisure encourages increased driving. If any uncertainty remained in the picture, it was whether or not increments in demand would exceed the 5 to 7 percent growth rate since 1970. In any case, there was no chance that they would be less.

In the face of such expansive growth in demand, what do we find? First, there has been a moratorium on refinery construction and expansion within the U.S. Only one new major refinery was built in 1972. None has been built on the East Coast for fifteen years. Between January 1961 and January 1971, U.S. refinery capacity has been increased only 24.6 percent—less than half as much as demand has increased. In 1972, while demand was increasing by 7 percent, capacity grew only 2.2 percent. Demand was clearly outstripping productive refining capacity. It was only a matter of time before a shortage would occur.

Apart from the embargo, several factors working together brought the matter to a head in 1973, not 1974 or 1975.

First, there was an especially heavy demand for heating oil in the winter of 1972. This prevented the off-season build-up of gasoline supplies to meet the expected 1973 summer demand. Consequently gasoline inventories were 10.6 percent lower on April 1, 1973, while the demand was 6.1 percent higher. But, even without the 10 percent drop in inventories from 1972 levels, the 6 percent increase would have strained capacity and, unless new capacity were placed on stream late in 1972, a 1973 shortage was almost sure.

It is true that the trade-off between the production of heating oil and gasoline complicates the planning problem. Still, the rising demand for heating oil was as predictable as the rising demand for gasoline. Shortages of natural gas supplies and governmental rules regulating the combustion of high-sulfur coal and residual oil caused many industrial users and utilities to shift to heating oil. This shift, coupled with an especially bad winter in the upper Midwest, placed new demands on available refining capacity, but they were foreseen and were not likely to abate. One wonders why this additional demand for a refined product which is both substantial and permanent was not translated into new refining capacity. Not only would this increase in demand be regarded as an opportunity by a competitive industry, but it would be seen as an ideal increase since the increase in the demand for one seasonal product (gasoline) was balanced almost perfectly by an increase in demand for a counter-seasonal product (heating oil). This assures balanced refinery utilization. Perhaps the *Oil & Gas Journal* was right when it theorized in the March 27, 1972, issue:

> The conviction of many refiners—who point to widespread price-warring and distress gasoline—is that a lull in construction is needed to let demand catch up a bit. That sounds reasonable—if the lull doesn't last too long. At this point there is a definite danger of this.

Two factors add credibility to this theory:

1. Refining capacity in the early 1960's had been held to be in excess from the majors' standpoint in that more gasoline was being produced than could be marketed through their own controlled outlets operating under national brand names. This resulted in a general weakening of prices and a rapid growth of independent marketers who could buy this additional gasoline at favorable prices.

2. As soon as it became clear that the shortage seemed to be producing a negative public reaction which might backfire on the industry, over two million barrels per day of new refining capacity was announced. Most of this new capacity will be achieved through the expansion of existing refineries. This step could have been taken at any time in the past since it involved no controversy over sites.

Oddly enough, at the very time refinery construction was being curtailed in the U.S., the same major oil companies were on a virtual binge of refinery construction in other countries around the world, particularly in the Caribbean area. It appears from all these facts that the gasoline shortage is a U.S. problem, not a worldwide problem.

From the mid-1960's on, fewer than a half-dozen refineries in excess of 50,000 barrels per day (b/d) were built in the U.S. Yet the story overseas is quite different. The list of refineries in the Eastern Hemisphere (on page 12) is instructive.

The Interior Department's Office of Oil and Gas has estimated that 1,720,000 b/d of refining capacity has been "exported" to foreign locations.[1]

Company	Location	Capacity in B/D
Socal	Fehry, Belgium	130,000
	Osaka, Japan	76,000
Exxon	Singapore, Malaysia	81,000
	Okinawa	72,000
Texaco	Eastern Hemisphere	918,000
Mobil	Amsterdam	80,000
	Woerth, Germany	70,000
	Wilhelmshaven, Germany	160,000
Gulf	Europe and Far East	166,000

—(*Philadelphia Inquirer,* July 22, 1973)

The problem of rapidly growing demand in the face of static refining capacity was further aggravated by still another, more technical problem. Two-thirds of U.S. refining capacity—including almost all of the independent capacity—is incapable of handling high-sulfur crude. Most refineries were originally designed to operate on domestic, low-sulfur crude oil. Since April 1972, however, domestic crude has been in very short supply. "Allowables" under state production controls have been raised to 100 percent. Because of the domestic shortage many refineries—especially the Midwestern independents—are operating at less than capacity levels. Not only does this accentuate the shortage, but it focuses the harm of it on their customers, mainly the independent marketers. This in turn effectively reduces price competition at retail. The end result is that the price of gasoline rose more during 1973 than during the previous twenty years.

What Should Be Done about the Gasoline Shortage?

There may be a shortage of gasoline, but there is certainly no shortage of proposals for dealing with the shortage—recommendations abound. Some are quite short-run and are concerned with preventing the shortage from being used to eliminate competition in the industry. Others are long-run and are directed at a fundamental restructuring of the industry. Some are concerned with changing driving habits and life styles. Others are directed at changing the structure of incentives that influence the behavior of buyers and sellers in this business. Some are directed primarily at private decision makers, others at federal and state officials. Some involve questions of economic policy, others are fundamentally technological. All need to be taken seriously and carefully evaluated.

Short-run Proposals

The primary short-run solution to the shortage of 1973 has been to try to run all refineries at full tilt with as high a yield of gasoline per barrel of crude input as possible. To this end, in those cases where independent refineries were unable to get adequate supplies of sweet crude, the government has directed its "royalty oil" into those refineries with some beneficial results. In addition, the government has adopted a voluntary allocation program, but that has had negligible results.

On the whole, the "full tilt" proposal worked reasonably well during the 1973 shortage. The magnitude of the short-fall remained at the nuisance level for the average motorist, although it became critical for the independent marketer with no in-house refining capacity.

A secondary short-run solution was to increase imports of gasoline from Western Europe. But this proved less effective. First, there was little extra gasoline in Europe that could be exported to the U.S. and, second, the price of this gasoline skyrocketed. The high level of tanker rates in 1973, when coupled with high product prices in Europe, made the price of imported gasoline unattractive. This was especially true for the independent marketers who couldn't pay the high price for imported gasoline and remain competitive.* Some of the cargoes landed from Holland and Italy were priced very near to the major brand retail price of gasoline, and this was before marketing costs were added. Finally, the failure of price controls to allow the averaging of product costs (high priced imported with low priced domestic) discouraged the importing of gasoline. Whether this will hold true, now that cost averaging is permitted, depends on the level of tanker rates and the supply of European gasoline. Normally the European refiner converts roughly 14 percent of his crude through-put to gasoline, as compared with the 45 percent or more that is typical of the U.S. refiner. Even so, there is some spare refining capacity in Europe. Hence, the future value of this solution to the U.S. gasoline shortage problem depends on how the European refiners react to the opportunity to export gasoline to the U.S. (Recall, if you will,

---

* Unbranded gasoline prices are quoted in numerous ways: cargo, tender, barge, wholesale rack, etc. The "rack" is the facility at a refinery where transports load.

that much of the European refining capacity is owned by American multinational oil companies.)

The most obvious way to curb the short-term demand for gasoline would be to let the price rise to a level that would reduce demand to the point where it just equalled supply. This is the traditional market solution to the problem of scarcity. Normally this would be the preferred solution. In this case, however, it poses several problems.

The first problem is that such a solution would merely result in massive windfall profits for the major oil companies without generating much, if any, additional supply.

The second problem is that gasoline is such a basic commodity in our everyday life style that it would take price increases on the order of 50 to 100 percent to curb its consumption significantly. This means that not only would the windfall profits be dysfunctional, but they would also be enormous. Without the free entry of potential new suppliers to bid the price back down to the marginal cost of producing gasoline, or without expansion of capacity by existing major suppliers and subsequent price competition to bring the price back down to the marginal cost, no benefit to society would result from allowing profits to rise to unprecedented levels. The structure of the industry makes the entry of new suppliers and price competition among the majors unlikely. Consequently, this solution is not especially attractive— except to the expectant beneficiaries of this proposal.

A related solution to the shortage problem would be to substantially increase the federal fuel tax. This would largely eliminate the windfall profit problem. One of the shortcomings of the tax proposal is that it would not stimulate supply, but would only contract demand. Second, the tax proposal is regressive and would fall relatively heavily on the working class. Nonessential consumption would be initially curbed, but commuting to work would still be required, whatever the price of gasoline. A ten-cent tax increase could mean as much as three dollars' per week increase in transportation to the average working man. Such a proposal is not likely, therefore, to have great political support, especially if it merely rations the shortage without relieving it.

Several additional ideas for reducing the demand for gasoline were proposed. Bills were introduced in the 93rd Congress, House of Representatives, to cut all speed limits in excess of 50 mph by 10 mph and to provide an additional itemized tax deduction for the use of a passenger automobile in a carpool.

In addition to propaganda campaigns encouraging self restraint and outlining better driving habits, proposals of a ten-hour, four-day work week and improved mass transit have also been advanced.

Long-run Proposals

To the extent that uncertainty about crude oil availability has delayed needed refinery construction and expansion, moves to increase crude oil availability can be seen as indirect measures for increasing refining capacity. Within days after the President announced elimination of all duties and abandonment of volumetric import quotas (in favor of an unlimited license-fee system), the majors began to announce significant expansions in refinery capacity. Such announcements now exceed two million barrels per day.

Recent Congressional action to promote the early construction of the Trans-Alaska pipeline and to remove legal and environmental roadblocks can also be seen in part as an indirect means of encouraging refinery expansion.

Also, proposals to promote more drilling on the outer continental shelf and to liberalize the depletion allowance in order to encourage greater domestic exploratory-developmental drilling and production are also indirectly aimed at promoting the expansion of refining capacity.

To further encourage refinery construction and expansion, the President's new Oil Import Program provides that refiners will be able to import crude oil without a fee for five years in amounts equal to 75 percent of the new capacity. As the scheduled fees for crude oil increase every six months from 10.5 cents per barrel in May 1973, to 21 cents in March 1975, this may be an increasingly attractive inducement.

Proposals to change the federal income tax laws, which now favor off-shore construction, so that they will promote domestic refinery construction, are being considered.

Also, to the extent that environmental restrictions on refinery siting and effluent have curbed refinery expansion and capacity operation, proposals to relax these standards could increase refinery capacity.

Breaking Economic Power

As noted earlier, one of the unsettling elements of the gasoline shortage is that for whatever reason—conspiracy, conditioned reflexes, misplaced prudence, or bad planning—a small group of powerful businessmen could, and did, decide to let gasoline demand catch up with refining capacity in the United States while expanding refining capacity at a breakneck rate over the rest of the world. This does not sit well with the average American's pragmatic notion of democratic pluralism. As a people we have always been uneasy with the visible exercise of private power. The gasoline shortage made the private power of the major international oil companies visible as it has never been visible before.

Consequently, while the largest number of proposals considered to date have been concerned with the amelioration of the inequities attributable to the present shortage, the most thought-provoking proposals have been directed toward the fundamental restructuring of the industry, so that future shortages, if any, will be the result of market decisions rather than private decisions. Such proposals are aimed at diffusing economic power that has become concentrated in a small handful of huge international oil companies who have been running this vital industry in their own interest for the last fifty years.

Several of the proposed structural remedies are short-term in their perspective and are aimed primarily at preserving some semblance of price competition in the industry by preventing the summary destruction of the independent, private brand marketers and refiners who would otherwise be forced to bear a disproportionate share of the burden of the shortage. In the spring of 1973, the Economic Stabilization Act of 1970 was amended by the so-called Eagleton Amendment, giving the President or his delegate the power to allocate petroleum products to marketers and crude oil to refiners. But this granted allocation authority was implemented only on a "voluntary" basis. Its purpose was to provide all previous customers with the same proportion of the supplying company's available supply as they had had during a specified base period. In a sense, it undertook to "share the shortage." But within thirty days of its inception, it was clear that the voluntary program was not working well. Something else was needed. Yet mandatory allocations were opposed by the White House, and a series of parliamentary moves prevented legislative action on a mandatory program before the Congressional summer recess. In the end, the House Rules Committee would not allow the mandatory allocations rider on the Housing Protection Act to get to the floor of the House because of another rider eliminating price controls on beef which would take the President off the hook with regard to beef shortages.

Some attempts were also made to preserve independent competition by granting import quotas to distressed operators and, as noted earlier, by directing government "royalty oil" to the independent refiners.

The most immediately significant feature of the gasoline shortage of 1973–74 is not the modest inconvenience caused to the motorist. Rather, it is the fact that the shortage has been carefully focused on the independent marketer and the independent refiner. Unless the government seriously undertakes to "share" this shortage, as well as to prevent future shortages, irreparable harm will be done to the chances for future price competition in the marketing of gasoline.* Without gasoline most of the independents will be forced out of business permanently. It is unlikely that they or other, equally aggressive, competitors will return after the shortage is over. The dependence of the independent on the willingness of his major integrated competitor for supply in time of shortage is a major weakness of the present structural arrangement. It is unthinkable that this arrangement should be allowed to continue. This is the primary rationale for basic structural reform.

The most obvious structural change would be to break the vertical linkages between crude production, refining, transportation, and marketing. If each major functional level of the presently integrated firms were forced to bid aggressively for its inputs and compete actively for buyers for its outputs, the chance of using a shortage to destroy nonintegrated or partially integrated competitors would be reduced.

Vertical divestment thus becomes a major reform possibility as an alternative to mandatory or voluntary allocations on the one hand and the demise of price competition on the other hand. It would

---

* A mandatory allocation program was not finally enacted until November 1973 and its implementation was deferred until late January 1974. A subsequent FTC evaluation of its first sixty days shows the administration of the act has been minimally effective.

also resolve the more fundamental problem of economic power. With each functional level divorced from the other, entry would be enhanced. If present firms decided not to expand capacity, or improve process or product, or introduce efficiencies, outsiders might find this reluctance to be their opportunity to profitably meet a need not adequately being served, and they could do so without having to enter all other levels of activity, as is now the case.

Several proposals* have been made to divest refining from marketing, or crude production from refining and transportation and marketing, or to divest each activity from every other. Such proposals are simple in form but complex in implication. Questions of efficiency, cost, investment incentives, and property rights are involved. In fact, the balance of this book will be concerned with developing the industry background necessary to make informed judgments about proposals for structural and behavioral reform.

In addition to divorcement, several other proposals have surfaced in the current search for workable structural reforms. Senator Jackson has proposed that the huge energy conglomerates be incorporated by the federal government (as opposed to state governments) and that they be subjected to detailed federal regulations. This proposal needs to be fully explored since it may be much less disruptive than divorcement.

The increasing injection of governmental authority in the international oil business makes nationalization a much more realistic alternative than it has ever been before. When governments buy and sell oil directly, private concerns are at a disadvantage. The international politics of oil is becoming crucial. Our foreign policy with respect to the volatile Middle East is involved. Our stance with respect to the development of the "Third World" and our concern for international stability and progress will be inextricably bound up in our oil policies. Dare we any longer entrust these sensitive and vital policies to the profit motivation of the major oil companies? If we use the power and influence of the United States to shape international oil policy, would it not be better if the industry at certain levels were nationalized so that no conflict could exist between the public interest and private interests?

The role of government in this business needs to be changed.

---

* Most of the recent antitrust suits against the "Big Eight" or the "Big Fifteen" have also sought divorcement as one of their remedies.

New forms of intervention and involvement need exploring. Present forms of intervention and involvement have been shown to be bankrupt. Where regulation has been effective, it has been wrong headed. The government has become a party to rigged prices, curbed production, and other arrangements which serve private interests more than the public interests in oil. This must cease.

Key Issues

Why has there been a gasoline and distillate shortage and what should be done about it? This is the central theme of this book. The answer to the "why" part of the question is necessary before we can intelligently deal with the "what should be done" part.

The petroleum industry has now reached a critical stage in its evolution. Public policies that have governed the development of the industry over the past three to four decades have begun to break down. Shortages of gasoline and other refined products now join the long list of energy problems. Gasoline, when consumed in the internal combustion engine, is a major source of air pollution and, hence, deeply related to matters of environmental reform. Annual per capita spending on gasoline is substantial. Therefore, small changes in gasoline prices have a major impact on household budgets and price level stability. The retailing of gasoline involves more than 220 thousand retail service stations. A considerable amount of prime real estate, unskilled labor, and specialized capital is at stake.

The importance of gasoline to our mobile society which is organized around the personal automobile is obvious. The efficiency with which this product is made available to the driving public is a matter of great social consequence. Waste or inefficiency, unreasonable prices, or unresponsive institutions which ride roughshod over the variability of consumer preferences cannot be tolerated.

Not only are the policy decisions which need to be made important, they are likely to be irreversible. If the decisions we make by design or default turn out to be the wrong decisions, we will not be able to start over again from scratch. Many key elements will no longer be present. It is crucial that we explore as carefully as possible the long-run implications of the options we face. Short-run, improvised solutions to any number of the so-called crises that are upon us could very easily be disastrous over time.

It is common these days to talk about "system effects." We recognize that policy "A" is related not only to goal "A" but also to goals "B" and "C." Something of this sort is involved in the area of gasoline and oil products. A decision designed to solve one problem may create a host of problems in related areas. It will be important to be sensitive to this possibility in the gasoline industry which is part of both the energy industry and the transportation industry. For example, decisions about lead additive restrictions to protect catalytic mufflers may substantially reduce retail competition in the marketing of gasoline if independent marketers are denied reasonable supplies of the new gasoline.

Sensible proposals for industry reform require some appreciation of the historical evolution of the industry. In most cases current institutions and arrangements did not spring full-blown into existence but followed as consequences of a long series of problems and solutions.

For example, there is no necessary reason why most gasoline should be sold through specialized retail outlets controlled by vertically integrated, giant corporations who produce and refine crude oil and sell the resulting products under nationally advertised brand names, often on credit. It may seem natural now because this arrangement is so pervasive. But it was not always so and it may not be the most efficient and responsive way to market gasoline, even today.

In the next several chapters we will examine the structural evolution of the gasoline industry and how the central marketing styles have developed. Following this we will look at the competitive developments over the last decade which led to the crisis during the summer of 1973. Once the historical context within which the shortage occurred is understood, the proposed reforms and their larger implications should be seen in a clearer light. We hope to shed such light by looking at the facts.

2

Competition, Conflict, and Cooperation in the Marketing of Gasoline

Basically three types of competitors engage in various forms of rivalry for gasoline buyers. The most important type in terms of sales are the so-called majors. These are the seventeen largest oil companies who rank among the world's largest and financially most powerful corporations. All are vertically integrated backward into crude production, transportation, and refining. All operate in a multitude of different geographic markets. Some even operate worldwide. This category includes such household names as Exxon, Texaco, Gulf, Shell, and Mobil.

Besides this group of seventeen major vertically integrated marketers, there are approximately twenty smaller integrated marketers who are characterized as semi-majors. This category includes such companies as Fina, Dergy, Kerr-McGee, and Murphy. Because of their smaller size and frequently unbalanced integration, their market behavior typically is quite different from their major rivals. More often it parallels the behavior of the third type of competitor, the so-called independent marketer who may be only partially integrated or nonintegrated and who concentrates primarily on the marketing of gasoline. This category includes such companies as Martin, Hudson, Site, Autotronics, and Certified.

Competition in any important gasoline market involves hundreds and even thousands of stations controlled by these three different types of sellers, each vying for a share of the available business. Each type of marketer makes a different competitive offering based upon variations in products and services sold, free services offered, quality of operation, and retail facilities, location, price level, method of payment, advertising support, and sales promotions. Each individual marketer has a unique strategy for his brand, but each also abides by the general requirements of his class. The particular situation dictates which features of the competitive offer will be stressed and which will be eschewed.

Intratype versus Intertype Competition

Competition, then, takes two forms—*intratype* competition and *intertype* competition. Intratype competition is the competition between companies of the same type, Shell and Texaco, for example, or Martin and Hudson. Intertype competition is the competition between different types, for example, between the Shells and the Martins. This distinction is fundamental to an understanding of the regulatory role that competition plays in this industry. Without intertype competition there would be no real, effective, and persistent retail price competition. Intertype competition represents the sole force for revitalization in this industry. This has been true in grocery and drug merchandising as well. It is true in general merchandising where the self-service discount store has fundamentally challenged the dominant department store which was the undisputed king in merchandising appliances, housewares, and clothing in times past.

Forces at work in intratype competition keep competitors within the same class operating in similar ways. Because of the basic operating similarities of intratype competitors, and because of the local nature of retail competition, one company cannot allow another competitor of the same sort to have a competitive edge for long. If one seller introduces a merchandising gimmick or new promotion, others must follow suit or else lose position. Their marketing styles are too similar for such marketing moves not to influence motorists who are otherwise indifferent in their preference for brands within the same competitive category.

This feature of intratype competition means that price differences are even less tolerable. Since they are immediately countered, they become self-defeating and thus are seldom used. Consequently intratype competition plays down price competition and plays up non-price forms of competition which are harder to counter. These non-price moves in turn are initiated and end up as permanent additions to marketing costs which all customers must bear. The ultimate result of intratype competition is the destruction of price competition and the encouragement of wasteful marketing practices which, when adopted by all, become embedded in the general cost of doing business.

The wasteful, expensive, homogenized, top-heavy marketing structure evolved by the major marketers is a direct product of this type of intratype competition. The financial commitment to their elaborate means of selling, in turn, generates an overpowering loyalty to the status quo and a strong appreciation of each competitor's mutual interdependence.

Intratype competition thus contains the seeds of its own destruction. As investment in more and more expensive marketing arrangements and practices grows, so does the sense of mutual interdependence. The futility of further competition becomes clear. The final conclusion of the process of intratype competition is cooperation. Cooperation is needed to keep the house of cards from falling. In industry after industry this is true—common circumstances and common fate breed common outlooks. The sense of all being in the same boat prevails; a spirit of cooperation and self-restraint emerges. Potential deviants are punished by the rest in the name of the common good.

Intertype competition is largely responsible for insuring that industries remain strong, efficient, and adaptive to the changing interest of the consuming public. By definition, intertype competition is juxtaposed to the status quo. Intertype competition by and large emphasizes differences. If intratype competition minimizes price competition, intertype competition can be expected to emphasize price. If one emphasizes low volume, full service operations, the other can be expected to challenge this marketing strategy with high volume, mass merchandising techniques and limited or self-service operations. If the natural consequence of the process of intratype competition is cooperation, the natural consequence of the process of intertype competition is conflict. This seems to be true in gasoline marketing just as it has been true in other industries.

As the major marketing style evolved and began to dominate, it spawned a counter style—that of the mass merchandising discount seller. The result has been twenty-five years of conflict with the majors defending the status quo by resorting to every weapon at their disposal and the discounters devising ever-new challenges to erode the majors' entrenched market share.

While the result has not been ideal, it has certainly been much better than if the majors were without a vital intertype challenge. After all, intertype competition is the only feasible way of keeping the lid on gasoline marketing margins. They are already scandalously high. Without the competition from the price marketers, they would have been twice as large. The experience of the recent shortage shows how quickly marketing margins will rise without the unrelenting pressure of intertype competition to keep them down—even when they are subject to federal price controls.

Intratype Competition at Work: Major Brand Marketing Style

The marketing strategy of the majors first began to develop into its current form in the early thirties. Before 1910 gasoline was sold by livery stables and garages and by hardware, grocery, and general stores. By the mid-1920's there emerged a clear need for a new type of outlet as a consequence of the dramatic growth of automobile registrations. The result was the development of the retail gasoline service station which was soon to dominate the retailing of gasoline.[1]

The initial demand for service stations was quickly met by thousands of independent businessmen who were eager to capitalize on the profit opportunity in this important new venture. The result was intense competition between the major oil companies for exclusive representation by the higher quality, independently owned retail outlets.

There were two consequences of this intense competition for quality representation. The first was that the price of quality representation became quite high. The other was that the frequent brand switching by the independently owned outlets often disrupted major oil company representation in an area.

For both these reasons and for others, the major oil companies began to integrate forward to control their key marketing facilities, either by outright ownership or by long-term leases.[2] Integration also assured that the outlets would sell only the brand of the owning company. It also permitted greater control over station appearance and operation. Such control is obviously very important in the development of a favorable brand image.

While the ownership and control of marketing properties permit employee operation, most major companies have chosen to use independent dealers to run their service stations. Under this arrangement the major oil company is both landlord and supplier.

This was not always so. Initially, it was quite common for the major oil companies to operate their own stations with company employees. For a number of reasons, including the avoidance of chain store taxes passed by the states in the Depression, most majors had moved away from direct station operation by the mid-1930's. The threat of unionization as well as the chance to shift the burden of low returns from retailing to others also entered into this decision.

Today, all but two majors use this approach. Calso and Sohio are the two exceptions, between five and thirty percent respectively. The others operate fewer than one percent of their outlets with company employees. For example, Gulf only operates 32 stations out of 29,540 and Shell only 171 out of 20,464 stations.[3]

Marketing Strategy

As the marketing structure shifted to the single brand, dealer operated, supplier-controlled service station, other elements of the majors' marketing strategy began to take shape.

First, the nature of the competitive setting dictated that *price* competition had to be minimized. With only a few major sellers in a given market, price cuts were self-defeating. If one seller cut price, all others had to follow suit or else lose market share. Since the motorist does not necessarily buy more gasoline as the price falls, the only effect is a loss of profit by each seller.

Thus, such a competitive situation shifts the burden of competition to measures other than price cuts. For example, market coverage becomes a major basis for competition. Other factors being equal, market share is closely related to the share of outlets in that market. This fact has resulted in wasteful overbuilding by each of the majors in an effort to hold or gain market share. Each seller recognizes this form of competition to be self-defeating since it often reduces the volume sold through a single station to unprofitable levels. And yet, each seller is compelled to continue adding stations in order to avoid loss of share.

Brand image and product differentiation also assume a primary function in the major's non-price marketing strategy. The reason is obvious. If price is not to be the basis for consumer preference, other bases must be developed. Many elements enter into the "image" that the driver forms of a particular brand. Some are psychological in character while others are aesthetic—station design, color schemes, and logo symbolism.

Brand image and accompanying brand preference is importantly shaped by advertising—especially television advertising. When a product is difficult to assess as is gasoline, slogans and symbols play a key role in the development of customer preference. "Put a Tiger in Your Tank" or "You Expect More from Standard—And You Get It" have been extremely successful image builders. Note, these campaigns say nothing about the product as such.

Other marketers seek to build a favorable brand image around a technical aspect of the product. Shell's special mileage ingredient "Platformate" or Mobil's "detergent" gasoline are good examples of this approach. Since most gasolines contain "platformate" or its equivalent, Shell's very successful campaign might be compared to a bakery advertising that its bread contains flour.[4]

The major oil companies have also found that the retail setting within which gasoline is sold is used by many motorists as a basis for forming an image of the unseen product. Uniformly clean, well run, modern appearing, quality controlled outlets imply in the minds of many motorists that the gasoline sold by such outlets is of uniformly good quality and thus motorists form a brand preference on this basis. For this reason, the major oil companies spend a considerable part of their marketing budget on modernization and rehabilitation of older stations and in designing and constructing modern, aesthetically appealing new stations. Also, considerable time and effort is spent in supervising station operations and appearance in order to foster a favorable image for their brand.

As important as brand image is in providing a basis for brand preference, strong preferences for a single brand are uncommon. In any given market, most motorists would regard the top half-dozen major brands close substitutes for each other. When price dare not be used, and when several competing brands have developed equally good images, other patronage techniques are required. Credit cards, stamps, short-term premiums, and games are widely used by major brand marketers.

Gasoline credit cards came into widespread use after World War II. Obviously, consumer credit is not needed to finance a five to seven dollar purchase. Credit is an out-and-out patronage continuity device designed to reduce brand switching. Trading stamps serve the same purpose. If a motorist participates in a stamp plan at her supermarket, she might also opt for a gasoline brand offering the same plan, other things being equal. Double and triple stamps are occasionally used as straight-out promotions on slack days, but primarily stamps promote patronage continuity.

Short-term continuity premiums are also used to induce some drivers to switch brands and to give a bonus to existing customers. Premiums such as steak knives and salad bowls have long been used. Games of chance were in vogue for three or four years during the late 1960's and were quite effective.

Market Structure

Conveniently located, modern, aesthetically appealing, supplier controlled, dealer operated, retail service stations offering a single nationally advertised brand of gasoline supported by a host of continuity promotions including gasoline credit cards had become the normal way for the major oil companies to market their gasoline by the 1960's. In part this approach represented a calculated decision by each major vertically integrated oil company to gain some measure of control over the demand for their most important product—gasoline. But it also represents a predictable response to key structural features of the industry—namely, vertical integration, price interdependence, and the cartelization of crude production.

As already noted, the industry is dominated by seventeen so-called major oil companies which produce, refine, transport, and market crude petroleum and refined products. These companies refine over 80 percent of all domestic gasoline and sell roughly 70 percent of all gasoline purchased through their controlled, branded out-lets. The remaining gasoline produced by the majors finds its way to market through unbranded and private brand channels.

When each seller's price affects the sales volume of the other sellers in the field, the sellers must assiduously avoid price competi-tion, since price cuts by any one will have to be met by the others. If, as a result of the lower price level, the total demand does not in-crease, each will merely sell the same volume at a lower gross mar-gin. For example, if a Texaco station cuts its price, that affects the sales of the Shell station across the street, and the Shell station will have to respond. This process explains why non-price competitive techniques dominate the majors' marketing style.

Vertical integration, on the other hand, explains why major brand refining and marketing operations need not be profitable in their own right. When firms are vertically integrated, transfer prices and, hence, gross margins at each level are set by managerial action rather than by market forces. The result is that the profit at a par-ticular level can be manipulated as the logic of the integrated opera-tion dictates. In the case of the major oil companies, this allows a top-heavy, expensive, and inefficient marketing structure to be main-tained by profits from other levels.

Cartelization and Preferential Tax
Treatment of Crude Production

At first blush, a strategy of subsidizing operations at one level with profits from another sounds foolish. But if profits at a particular level are not competitively determined, and if all competitors are not completely integrated, subsidization makes good competitive sense from the point of view of systemwide profits. The advantage of shift-ing profits from one level to another is increased when, as in the case of petroleum, special tax advantages accrue to crude oil profits and not to marketing or refining profits. Profit shifting also permits the integrated firm to regulate the competition and growth of its non-integrated rivals.

The cartelization of crude production in the context of vertical integration, however, is the key to understanding the way gasoline is marketed by the major oil companies. Without the ability to earn high tax sheltered profits from crude, and without the ability to shift profits among the levels of the vertically integrated firm, the high cost, inefficient marketing structure of the major brand marketers would collapse from its own weight.

Two characteristics of crude oil production heavily influence the way the majors market their gasoline. First, the price of crude oil is rigged, that is, the price of crude oil is openly administered by the major refiners. Independent producers have to take what the major refiners offer. But this is not so bad for them, since the major refiners are also the major producers, and hence they have an interest in high prices* These high crude prices are in turn supported by state regulatory authorities who, until recently, used the device of demand pro-rationing to keep crude oil production balanced with the amount of crude oil demanded at the administered price. In the name of conservation, individual producers of crude are denied the right to increase their output to their maximum efficient rate by cutting price. Since they can sell all they are permitted to produce at the rigged price, and since they can sell no more at a lower price, they obviously support the rigged price. The traditional market calculus by which prices are set through the interaction of supply and and demand is thus effectively frustrated.

The one loophole in this arrangement was the potential availability of plentiful supplies of lower priced foreign crude. As soon as this threat became a reality in the late 1950's, however, the major oil companies were able to convince the government that rising imports of crude from abroad would threaten the domestic crude industry and hence increase our dependence on foreign crude. And since foreign crude is less dependable than domestic crude in times of hostility, quotas were set to regulate the importation of foreign crude in the name of national security.

This cartel, abetted by the state and federal government in the name of conservation and national security, is only half of the picture. The other half is the preferential tax treatment of profits from

---

* Any refiner who is 81 percent self-sufficient in his crude requirements finds it advantageous to take his profits at the crude as opposed to the refining level. If part of the increased crude cost can be passed on by the refining level, the required self-sufficiency ratio drops dramatically. For example, if 20 percent of the cost increase is passed on, the required ratio is .65. This figure embraces most of the majors.

crude production. The latter explains why the major oil companies want to shift their profits to the crude level; the former tells how they are able to effect this shift.

There are three major elements in the preferential tax treatment afforded crude oil profits—percentage depletion, expensing options, and foreign tax credits. The effect of these measures individually and in concert is to reduce the federal income tax rate on profits earned in the production of crude oil to a fraction of the rate at which profits from refining and marketing would pay.

The Necessity of Downstream Control

The logic of the total, vertically integrated system and the way it affects marketing is now obvious. For the system to maximize profits, high crude prices must be set and as much crude as possible produced at that price in order to earn crude profits and to utilize the tax preferences. In order to protect these profits, the large crude producer now needs to control its crude market. This means it must have adequate refining capacity so that it has a high priced guaranteed market for its crude. If most refiners were independent of crude production, they would have an interest in lower crude prices and hence would force down the high administered crude oil price. Controlled refining is thus essential—not for its own sake—but because it is necessary to support high crude prices.

The same is true of the need for controlled marketing. Unless a major producer-refiner can sell a large share of its gasoline through controlled branded outlets, it would be required to sell its gasoline to marketers who would play refiners against each other and thus weaken the price structure all the way back to crude.

Controlled refining and marketing is thus essential to the maintenance of high crude profits. Further, by holding refining and marketing margins low relative to crude margins, the growth and behavior of the nonintegrated and partially integrated refiners and marketers can be carefully controlled. Thus does market structure affect the marketing style of the major oil companies. The price interdependence of the major sellers reduces the role of price competition and justifies the emphasis given to non-price methods of competition. Vertical integration in the context of a noncompetitive, tax favored, crude oil market places marketing in a supportive role. Inefficient, costly marketing arrangements make sense because they sustain and enlarge high crude profits.

Intratype competition in the context of such a marketing structure has produced a Frankenstein monster which neither the sellers nor the buyers want. The sellers are afraid to change unilaterally. The buyers figure they might as well take the marketing excesses since they are obliged to pay for them whether they want them or not. Such a situation creates its own counter-strategy when intertype competition is free to function. The price marketer's strategy is best understood as a counter-strategy to the major strategy. It appeals to the very preferences the major strategy neglects. Intertype competition, when it is allowed to function, effectively checks the excesses of intratype competition.

Intertype Competition at Work: Price Marketers

While the modern day price marketer had his origins in the "trackside" stations and cut-rate operations of the Great Depression, his importance as a significant counter-strategy and intertype competitor is a post-World War II phenomenon. Several factors contributed to his development. In particular, the massive population shift to the suburbs and the resulting growth of automobile commuting created a key market opportunity that was ready-made for the price marketers.

The price marketers who emerged in the late 1940's and the early 1950's were different from the pioneers of the 1930's. The postwar strategy was to build large, modern, attractive, multipump stations on major traffic arteries between the suburbs and the central city in order to intercept increasing commuter traffic. Yet, in spite of the prominent role given large scale modern facilities, price remained the determining variable. So long as intratype competition between the dominant major brands' marketing style played down price, the principal counter-strategy of the intertype competitor was bound to emphasize it. It represented a direct appeal to an important segment of the gasoline market that preferred lower priced, streamlined operations to the higher priced, full-service convenience operations.

Economics of Mass Retailing

There is nothing novel about the marketing strategy developed by the price marketers: basically low price, high volume, and low unit cost. It is essentially the same strategy followed by other mass merchandisers such as supermarkets and discount stores.

When analyzing the modern price marketer's strategy, it is impossible to say which comes first—low price, low unit cost, or high volume. These three elements complement and reinforce one another. Low unit cost makes low price feasible. But unit cost depends on high volume, and volume is largely a function of relative prices.

There are several ways in which the price marketers affect savings which permit them to sell profitably for less. First, the price marketers' gasoline costs less since it is unbranded. Branded gasoline costs more because of the costs of product differentiation and brand promotion—credit cards and the rest. In normal periods, this represents a savings of one to two cents per gallon. It can amount to more when the supply of unbranded gasoline is greater than the prevailing demand.

Second, the price marketer often has lower fixed costs than the major brand marketer, since he typically avoids the extremely expensive corner location. Also, since most price marketers do not offer repair service, the physical properties are simpler and less expensive.

Savings in operating costs follow from the relatively high station volume of the price marketer. Because of this high volume, fixed operating costs are spread over more gallons and thus the cost per gallon drops. Even variable operating costs do not increase proportionately to volume. Consequently, they are spread over more gallons and hence are lower on a per gallon basis.

The cost savings that the price marketers have been able to effect give many of them a total operating cost advantage including profit of 4.0 to 5.5¢ per gallon. It is because of this cost advantage that the price marketer is profitably able to undersell the major brand retailer by several cents per gallon. On occasion this cost spread can rise to 7.0¢ or 8.0¢ per gallon.

There is no normal price spread between the major station and the price marketer. They will be greatest during periods of peace and narrowest during price wars. They range from 5¢ to 2¢ or 3¢ per gallon depending on the market and the nature of the competition.

In contrast with the major brand operation, most price marketer stations are company operated rather than dealer operated. The operation of high volume, low-price stations requires quick adjustment to competitive developments. This is difficult to accomplish under a dealer arrangement. It would also be difficult to regulate the earnings of a dealer in such an operation. He would make too much when prices were up and too little when prices were down.

The basic differences between the typical price marketer and the major brand retailer are summarized below:

Price Marketer	*Major Brand Retailer*
Emphasis on lower price	Emphasis on advertised brands
High station gallonage	Lower station gallonage
Few centrally located stations	Many conveniently located stations
Company operated stations	Dealer operated stations
No repair work, little TBA (tires, batteries, accessories)	Complete car service and TBA sales

Note that each of the price marketer's emphases counter the major marketing style. Thus, they provide a meaningful alternative to the majors' offer in all important aspects. Without such counter offers the motorist would have only a meaningless choice between different brand images associated with almost identical major-owned offers.

Semi-Majors: Hybrid Price Marketers

The third class of actors in the gasoline drama are the "semi-majors." They market like the price marketer by emphasizing price as their major purchase proposition. In other respects they resemble the major oil companies, since they are vertically integrated and use dealers and jobbers rather than direct operations. In terms of size and operating scale, however, they are dwarfed by the majors. Their marketing strategy reflects this difference in size.

By and large, the semi-majors represent a passive competitive force. Their whole style is nonaggressive. They are too small to afford the massive advertising investment required to develop a competitive brand image and too large to get away with being aggressive price cutters. They accept lower prices by default rather than by decision. Like the secondary major brands,* semi-majors merely serve to insulate rather than to threaten the major's excessive retail margins. They would be delighted to sit a penny or two under the major brand price and take whatever business comes their way. The higher the price, the better they like it.

To the extent that semi-majors do play a role in regulating market behavior, it is in their sale of unbranded gasoline to the independent price marketers rather than in the sale of gasoline under their own brand. Next to the independent refiners, the semi-majors are the principal source of unbranded gasoline. Furthermore, many of the majors who refuse to sell to the price marketers will sell to the semi-majors. For this reason, the semi-majors, in turn, are able to divert a sizeable portion of their own output of gasoline to the unbranded market.

Conclusion

The marketing structure of the gasoline industry dictates that the dominant marketing style of the major marketers emphasize non-price competition. The major integrated oil companies cannot and will not, as a general rule, compete with one another by discounting prices. Apart from the fact that price cuts in such a structure are self-defeating, the major integrated companies gain from cooperation with one another in a variety of ways including the joint exploration for crude oil, the building of pipelines, and the exchange of refined product. For them to suddenly become very competitive in the marketing of gasoline on a price basis would be inconsistent.

On the other hand, the independent price marketers operate in the void created by the majors' reliance on non-price methods of competition. Independents do not cut prices because they are public spirited; it is profitable to do so. They are independent enough to try discounting, and they are small enough to get away with it.

Consequently, the price marketers serve a competitive function in the gasoline business which is completely disproportionate to the share of the market they command. The actual and potential gaso-

---

* These are discount brands owned by the majors as a means of participating in the growing price market for gasoline. Such brands as Sello (Mobil), Alert (Exxon), Freeway (Sun), Rocket (Arco), etc., commonly masquerade as independent marketers.

line volume that the price marketers drain away from the majors acts as a check on how far the majors can go with the non-price method of marketing gasoline and how high they can set retail gasoline margins. Because the independent price marketer both must and is willing to reduce price to compete, he plays a key role in regulating this industry's price.

Of equal importance to the independent price marketers' role of price regulators is their role as marketing innovators. They have no stake in the status quo. By virtue of being "outsiders" they are motivated to challenge the dominant way of marketing gasoline. Consequently, they provide the alternative offers which are essential in keeping the industry responsive to changing consumer preferences for new purchase combinations. Without the presence of a small cadre of effective, independent price marketers, the motorist would have no alternative to the homogeneous, high-priced, full service offerings of the major, vertically integrated oil companies that dominate the marketing of gasoline. The independent price marketers provide the only hope for essential intertype competition to keep the industry on its mettle.

3

The Majors' Initial Response
to Intertype Competition: 1960–1970

In the period following World War II, and especially in the late 1950's, the share of market held by the price marketers grew dramatically. As a consequence, intertype competition became intense. While comprehensive figures on the growth of the price segment of the business are not generally available, fragmented information very definitely shows the trend. For example, surveys made by Sun Oil Company in 1954 and 1960 showed that the number of private brand stations increased by 67 percent—from 4,755 to 7,941 in areas where Sun marketed.[1] In the Midwest and Southeast, where they had adequate access to unbranded gasoline, price marketers reached a 30 percent market share.[2] In Norfolk, Virginia, the share was 25 percent, while Indianapolis marketers had 35 percent, Minneapolis 42 percent, and San Antonio marketers reached an incredible 45 percent.[3]

This dramatic growth of the price marketers obviously posed a basic threat to the majors' marketing style. Every gallon siphoned away from the majors' outlets served to increase their unit cost since there would be fewer gallons to absorb the fixed burden. This challenge was not new. As noted earlier, it had its parallel in other industries. The supermarket and chain store posed a fundamental threat to the small, independent neighborhood grocery store and its wholesale supplier. Similarly, the discount store represented a basic challenge to the once-dominant department store in the general merchandise line.

Just as the independent grocery store and wholesalers struck back with legal harassments and economic power plays, so did the major oil companies. These old-line operators used chain store taxes, blue laws, retail price maintenance, and the Robinson-Patman Act, as well as actual or threatened boycotts of manufacturers to slow the growth of their new competition. In the case of the major oil companies, they resorted to destructive price wars, dealer subsidies euphemistically referred to as "price support" plans, fighting brands, buy-outs, and supply curbs. In most markets where price marketers were important and posed a threat to the majors' style, each of these responses was utilized from time to time.

The major oil companies laid the groundwork for coping with the price marketers in the latter 1950's, and they still use many of the techniques today.

Price Wars and Price Protection

For many years the traditional price differential between majors and price marketers had been about two cents per gallon With the latter's rapid growth, this prevailing differential increasingly became a subject of contention. The need to narrow the spread was argued in repeated speeches by major oil company executives. But not until 1961 was the first broad-scale enforcement of a one-cent differential undertaken. In mid-February 1961 Shell initiated its now famous one-cent plan in California, a move that was to ignite a series of devastating price wars through large sections of the country for four years or longer. It started when Shell lowered its prices to within one cent of the private brand operations recently purchased by other majors, such as Gulf's Wilshire brand, Continental's Douglas brand, and Signal Oil and Gas' Hancock and Regal brands. These major oil company secondary brands typically were priced even with the prices of independent private brand operations. This meant that Shell was pricing within one cent of most of the independent private brands. As Shell lowered its prices to within one cent of the private brands, the independents would move to restore the spread back to two cents per gallon. Within days, the price of gasoline would fall six to eight cents per gallon. As a result of the months of price warring

that ensued, the marketing losses for the independents as well as
Shell and the other majors were enormous.

The West Coast experiment was closely watched. If it worked, it
could be expected to spread to other parts of the country where price
marketers had made important inroads. Success with the one-cent
policy would mean that the counter-strategy could be effectively
contained. With only a penny differential, much of the appeal of the
independent private brand marketer disappeared. By 1962, it was
clear that the one-cent plan was working and that the challenge of
the price marketers had been effectively checked.

In March 1962 a survey of some 2,700 stations in greater Los
Angeles showed that most major brands' prices were only a penny
above the independents. In July 1963 a survey of 2,377 major brand
stations showed 95.7 percent at 30.9¢ while 87.2 percent of the 514
price marketers surveyed were at 29.9¢.[4]

As Shell's success with its one-cent policy became clear, other
majors picked up the idea. For example, in Wichita, where the price
marketers had half of the market, Texaco initiated an experimental
one-cent policy. On the day after Christmas 1962, Texaco narrowed
the spread to a penny. This set off a wave of see-saw moves which
caused prices to drop from seven to ten cents below normal in a
matter of weeks. There were 18 such downward spirals over the next
two and a half years.

The "Wichita story" was repeated throughout the midcontinent
and other areas where price marketers were strong, as the other
majors implemented their own one-cent plans in the early 1960's.
By 1964, for example, Phillips was reported to be enforcing a one-
cent policy in two hundred markets.[5]

Those competitors caught in the turmoil of the penny differential
wars were severely injured. This includes not only the price marketers and the independent refiners who sold to them, but also certain regionally concentrated majors and thousands of other major
brand dealers and jobbers whose margins were held to stop-out levels
for months at a time.

The one-cent differential and the resulting price wars stopped
the growth of the price marketers, without question. Market share
and station share were down relative to the shares the price marketers had held before the wars began.

Perhaps even more important, it forced the counter-style to

deviate from its primary discount emphasis and to adopt many of the costly merchandising gimmicks of its intertype competitors. Trading stamps, long hours, improved station appearance, price leaders and other non-price methods of competition were adopted to compensate for the disappearing price advantage.

The pricing problems of the early 1960's not only resulted in a general slowdown of the aggressive price marketers, but, in the case of some, it resulted in either dissolution or forced sale and conversion to major brand operations.

The impact was equally devastating for the regional majors such as Union, Richfield, Tidewater, and others. Both Union and Richfield sought to reduce their future vulnerability through increasing their regional diversity by merging. Tidewater withdrew from the West Coast altogether by selling its properties, first to Exxon and then, after this was blocked by the Justice Department, to Phillips.

Shell, on the other hand, lifted its West Coast market share from 11.7 percent in 1960 to 16.3 percent in 1969. Through its geographic and functional diversity and because of its size and financial power, Shell and certain other majors could "buy" increased market share. They could check or destroy intertype competitors in one geographic area with earnings from other areas or other functional levels; in the process they could still maintain satisfactory overall profit performance. The regional or functional specialist caught in the economic warfare among the giants was severely squeezed. In such cases "might" becomes "right," and operating efficiency is thwarted by financial power—an unhappy consequence for an enterprise economy!

Coincident with the spread of the one-cent differential price wars was a "tane price war." These sources of price disturbance resulted in a virtual marketing "blood bath" from 1962 until early 1965. The big thrust into low octane gasoline sales grew out of Gulf's unfavorable experience in the late fifties with its effort to market a superpremium grade of gasoline called "Gulf Crest."

Gulf announced in November that it was replacing Gulf Crest with a subregular grade of gasoline called "Gulftane." Its avowed purpose was to go after the growing independent market, where regular grade gasoline was typically being sold for two cents less than the major brand regular. Gulftane was ultimately sold in 36,000 stations in 39 states. Similarly, in January 1962 Sun announced plans to sell a subregular grade, Sunoco 190, in 8,700 stations in 24 states.

The majors and the price marketers found themselves on a col-
lision course over their respective responses to Gulftane and Sunoco
190. The other majors indicated they would not give more than a
penny differential to another major brand marketer—even for a
subregular grade. For similar reasons, the price marketers felt that
they needed a penny under Gulftane if they were going to survive.
Gulf's plan, however, was to meet the price marketers' price head-on.
This impasse set the stage for massive price wars.

As the "tanes" rolled into more and more markets, both the
other majors and the price marketers fought bitterly over the differential. The Midwest and Southwest were hardest hit. In St. Louis
and Detroit prices fell by eight to ten cents per gallon. The results
were devastating. Even Gulf began to see the problems with its
insistence on meeting the price marketer's price head-on. After
nearly two years of market chaos and several abortive attempts to
restore prices to normal levels, Gulf finally decided to price Gulftane
one cent above the independents' regular price. In those markets
where the majors were unwilling to give the price marketers more
than a penny differential, even this concession did not help much.
Some markets did settle down, but others remained chaotic until early
1965.

The private brand discount marketers were hard-hit by the "one-
cent" and "tane" price wars. In San Antonio, where the independents
had operated 20 percent of the stations doing 45 percent of the gasoline business before the price wars, private brand market share
dropped dramatically as many stations were sold to majors or closed.
The same general pattern emerged throughout much of the country
where independent private brand marketing had been strong.

Both the one-cent price wars and the tane price wars were
financed by intricate price subsidy programs of the major oil companies. These programs permitted the majors to focus their great
financial power against their smaller, partially integrated or nonintegrated competitors. Without price protection, a major brand
dealer would be unlikely to engage in extended price competition.
If a dealer operating on a normal margin of six to seven cents per
gallon were to cut price by three cents, he would be giving up 50
percent of his operating margin. On the other hand, if the major supplier were to absorb 70 percent of this three-cent price cut, the reduction in operating margin to the dealer would be only nine-tenths

of a cent. With such price assistance, dealers could be induced to decrease and increase their prices at the will of their major oil company supplier.

The major oil companies' price protection program permitted profits to be drawn from less competitive marketing areas and less competitive operating levels of the integrated company, and it allowed them to finance price battles in selected markets. During the price wars of the early 1960's, hundreds of millions of dollars went into price war subsidy programs each year. Most of the nonintegrated or regionally concentrated marketers were especially vulnerable to this focused use of overwhelming financial power. Consequently, their ranks thinned and their market share dwindled.

Price protection is granted to dealers in two different ways: (1) through temporary reductions in dealer tank-wagon price (the price a dealer pays for gasoline) or (2) through zone support. The former involves giving the same level of price protection throughout a broad trade area—Denver, for example. Temporary reductions are used to drop prices in cities and areas of the country where price marketers are a threat. Zone support is more refined. Many companies will carve a city or area up into hundreds of smaller zones. This permits lower prices to be programmed into only those parts of a city or area where price marketers are stong, while maintaining higher prices in other areas of the city where the price marketers are thinly represented.

The Great Price Restoration of 1965

For more than three years in the early 1960's, the majors went after the price marketers using the techniques of selective price wars. The strategy was effective and major brand market shares were improved. Many price marketers were eliminated, converted to controlled major brands, "house broken," or bought out. Those that remained were often located out of the worst price war areas, were particularly strong financially, or gained strength through geographical diversity.

By early 1965 it was clear that this blunderbuss approach had run its course. The time had come to try to make money from marketing gasoline as well as from crude production and other operations. Even though most of the majors showed adequate profit growth during the turbulent price war years, gasoline marketing had made no contribution. The burden of price support for billions of gallons of gasoline sold several cents below normal reference prices represented a massive drain, even for the major firms. The opportunity loss was tremendous. More importantly, the profits of the two largest international oil companies, Texaco and Exxon, had shown distinct signs of weakening in the last half of 1964—a reflection of weakening international markets for crude oil and refined products. Something had to be done. With soft international markets, there was now good reason to put the domestic market on a better paying basis.

Texaco—the only fifty-state marketer and one of the largest sellers of gasoline—led a nationwide restoration of prices to normal on March 17, 1965. What Texaco did was merely to withdraw its price support to its dealers across the country. This move was quickly followed by similar announcements from Exxon, American, and Mobil. As this occurred, the battle over the one-cent differential was quickly forgotten and the price spread between the majors' and private marketers' prices for regular grade gasoline returned to two or more cents per gallon. While periodic price disputes erupted in specific markets, gasoline prices remained generally strong for the next four years, that is, until the end of 1968.

The results were predictable. With the counter-style free once again to use price differentials as a primary competitive instrument, the price marketer's share again began to grow along with his profits. It was an uneasy peace, however, and it could not persist.

The Search for New Responses

The vigorous revival of the counter-strategy during this interim period posed a number of serious problems for the majors caught up in an inefficient and costly marketing style. Old responses would not work. The new breed of price marketers—those who had come through the severe price wars intact—was to a degree immune to the old blunderbuss approach. Those who had survived the early blood-bath had learned their lesson. Most limited the number of stations operated in any single market so that even a marketwide price war only affected a small fraction of their outlets. Consequently, the majors found it difficult to focus their financial strength on selected independents without huge costs. They had to subsidize hundreds of stations in order to compete with a few dozen scattered price marketers. Also, the broader merchandise base of some of the newer independents gave them greater stability of operations that made them less subject to the pressures of price wars. In addition, the profits from crude production were under continuing pressure, which made retail subsidization even less attractive. It could no longer be done now without a visible, negative effect on their published financial statements.

Several new responses which sought to contain the growth of the counter-style were tried. Efforts at further refining zone price supports were tried. Cities such as Chicago and Los Angeles were divided into literally hundreds of small geographic zones. When an especially effective price marketer began to draw big volume, deep price support would be introduced in that zone and feathered out gradually in surrounding zones in an effort to focus the price subsidy without incurring a Robinson-Patman violation.*

While zone pricing undoubtedly slowed the growth of discount gasoline marketing, it did not curb the growth of the price marketers. The problem with zone pricing was that it was like a growing cancer that spread over time into previously unaffected areas, eventually enveloping the entire marketing area. The result was an intolerably heavy price support burden often involving extensive subsidization of literally hundreds of major stations in order to check a few price marketers. It was too big a price to pay. More efficient means were needed.

A second response to the rapid growth of the new breed of price marketers was an increased emphasis on non-price means of competition. If price marketers were largely immune from zoned price competition, then the majors would try to out-market them. The latter half of the 1960's became a period of marketing excesses.

* The Robinson-Patman Act makes it illegal to discriminate in the price charged competitors where the effect of such discrimination might substantially lessen competition. In a typical case, Shell dealers near an effective price marketer would be charged less for Shell gasoline than a Shell dealer in neighboring zones. If this favoritism significantly injured competition between these nearby Shell dealers, a Robinson-Patman violation might arise. By "feathering" the price cut out from the low-zone with small price increments, the oil companies hoped and argued that there is no significant competitive harm done. (See *Bargain Car Wash v. American Oil Co.*) Based upon the small size of the zones and the large arteries cutting through several zones which carry passengers back and forth daily, it is very likely that there is no way to prevent some dealers from being harmed by the inherently discriminating zone pricing approach.

Advertising budgets were dramatically increased. In 1968 Shell spent $26 million, Gulf $19 million, Exxon $17 million, Indiana Standard $16 million, Mobil $12 million, Texaco $12 million, Phillips and Sun 9 million each, and Socal $8.5 million.[6] Heavy investments were made in the construction of fancier retail facilities. Older stations were rehabilitated. Porcelain boxes were transformed into Cape Cod bungalows and sprawling ranch houses. Colonial brick facades and mansard roofs abounded. Canopies came down and colonial lanterns and arborvitae went up. Clean rest rooms were as important as clean gas.

This was also the period of "super-service." The service station truly tried to become just that. New stations incorporated multiple service bays. Advertising slogans promised quick and trustworthy service. The neighborhood dealer was "friendly" and "reliable" and concerned with your car's complete well being—not just selling gasoline. By forgoing the lure of low price, you could "contract" for your car's devoted care. Trading with your neighborhood dealer was like establishing a credit rating with your banker.

As merchandising gimmicks ran their course and became widely copied, others were introduced. Credit cards flooded the market in massive, unsolicited mailings until the practice was curbed by the Federal Trade Commission. Trading stamps and short-term premiums were proffered at every corner. Discounted soft drinks and sets of salad bowls or steak knives were featured in dramatic point-of-purchase displays.

Two extremes of the period were the games of chance and the free car washes. For a period during the late 1960's, gasoline and even clean restrooms were overshadowed by "Sunny Dollars," "Mr. President," and "Super Bowl." A trip to the station for gasoline was like a trip to Las Vegas. The success of the games in building traffic is well documented. But like other promotions, when all competitors offer them, they are neutralized and only add to the cost of marketing. The game spree was replaced by the car-wash binge, and like the game, the free car wash with each fill-up proved to be a fantastic traffic builder. Many stations were able to temporarily double or triple their volume at full reference prices by this device. But also, like the game, as it became widespread, it ceased to have such drawing power and tended to add more to cost than to revenue.

Some estimates place the cost of these promotions at one and one-half cents per gallon for an average fill-up of twelve gallons.

During this period the price marketers pretty much held the line on merchandising gimmicks. They maintained or increased their discount. Some got into trading stamps, cash redemption stamps, and continuity premiums. But in most cases these promotions were used like additional discounts rather than to promote patronage continuity. More often than not, they amounted to an additional, hidden price concession which in some cases amounted to two or three cents per gallon. Yet because they were more ambiguous, they appeared less aggressive than an additional two-cent price cut on the pump which might provoke major brand retaliation.

At the same time that many price marketers were using quasi-price cuts to increase the spread from major brand prices, they began to experiment with less service rather than more service. The "gasser" operation grew. In some instances a hermaphrodite self-service was instituted with traditional pumps and stations. (The evolution of self-service and its import on the economics of gasoline retailing is discussed in the appendix.) Again, their initial strategy was to differentiate the service in order to justify an additional discount rather than to introduce a new method of marketing. The emphasis was on increased volume through an increased discount rather than on lower cost and efficiency.

The two divergent strategies—service versus discount—had a profound impact on the comparative economics of gasoline retailing. The majors' strategy of increasing service added cost faster than it added volume. Higher cost facilities and expensive promotions with increased payroll made the majors even more vulnerable to the discounters who were able to decrease their marketing costs per gallon as a result of the increased volume generated by their discounts. More than anything else, the fact that the marketing costs per gallon of the two segments were moving in opposite directions signaled the coming end of the old competitive order. The emergence of full-fledged self-service and the tightening of profits from crude and international operations in the early 1970's would bring this old order to an end.

Zoned price support was too expensive and increased service and marketing tended merely to increase the major brand marketer's vulnerability—not his volume. In any event neither approach was effective in checking the price marketer's rapid growth during this interim period. The "gas-for-less" segment tended to become a recognized and permanent feature of the gasoline market. It could no longer be ignored by the majors since it had been accepted by the motoring public. Those giants of the industry that had fought the discount brand so bitterly a few years earlier responded by direct entry into this market through secondary brands such as Alert, Ride, Sello, Black Jack, PDQ, Jiffy, and others.

Periodic attempts were also made to raise the marketing margin by several abortive increases in dealer tank-wagon prices in 1969 and 1970 (See Chapter 4, Fig. 4–4). In each case these increases merely triggered new price wars and increased the burden of price support as the independents held their price and let the prevailing spread increase. Several of the majors who were short on crude relative to their product sales found this burden of widespread price support eating into their earnings. Continental and Phillips and Cities Service all experimented with terminal "rack" pricing plans designed to end the costly price war subsidy programs. In most cases, these pricing experiments failed when the big eight majors continued price subsidies to their dealers.

Throughout this period, low-priced, unbranded gasoline was readily available and much more of it was coming from the big eight since they were having trouble pushing gasoline through their terribly costly marketing system. Price discounting was growing and the large companies were fast losing their market position. The uneasy peace was coming unglued.

By late 1970, it was clear to the majors that something new had to be done. Profits were under pressure. Emulation was self-defeating, increased selling costs were making the majors' style noncompetitive, and price competition was too expensive. What to do? The only answers lay in either choking off the supply of unbranded gasoline or else in abandoning the old marketing style and going all out in switching to the new mass merchandising of gasoline. The stakes were fantastic. Over two hundred thousand traditional retail outlets and billions of dollars invested in brand imagery were involved. Obviously, the curbing of supply was the lesser of the two evils. How it was done remains yet to be disclosed. That it was done is a matter of public record.

4

The End of the Old Order

The breakdown of a long-standing industry practice can lead to a great deal of turmoil for companies as old ways are abandoned and new approaches are developed. Companies in the petroleum industry were experiencing such painful changes in the early 1970's. The "profit game plan" of the major petroleum companies was no longer workable and a new strategy was needed to rebuild the industry's profit structure.

Crude Oil Profit Cache

In the post-World War II period, the crude oil department of the integrated petroleum companies was established as the profit "nerve center" for the industry. The "locking in" of a disproportionately large share of profit at a single level of the integrated petroleum industry had long been the practice of the financially powerful oil companies.

At the turn of the century the Rockefeller trust had used its monopoly in refining (including key transportation) to keep crude oil prices low and to sell finished products at prices above a competitive level. In this manner the great Standard Oil Company both extracted monopoly profits and destroyed competition in the process.

The breakup of the Standard Oil refining monopoly in 1911 was followed by a shifting of profits in the petroleum industry backward to pipeline operations. The period of 1920 to 1940 was known as the "Golden Era of the Pipelines," at which time pipeline profits subsidized crude oil activities and downstream refining and marketing operations. The rate of return earned from pipelines far exceeded those obtained from other industry activities; at times it was the only source of profit to the petroleum companies.

By consent decree in 1942 the rate of return on appraised value of pipeline assets was set at 7 percent. This decree forced much of the excess profit from the noncompetitive pipeline operations which were largely owned and controlled by the major petroleum companies. Following World War II the profit cache for the petroleum industry moved further back in the industry structure to the crude oil division of the integrated petroleum companies. This was as far away as possible from the competitive marketing end of the business. The doubling of domestic crude oil prices from 1946 to 1948, and subsequent increases in crude oil prices in 1953 and 1957, moved the industry's profit center to crude oil production.[1]

Crude oil operations proved to be an excellent replacement for pipeline activities as the petroleum industry profit center for the following reasons:

1. The major integrated oil companies were able to administer the price of crude oil at the level they wanted because they not only produced most of the domestic and international oil but were also the principal buyers of the oil.

2. A combination of federal laws gave the integrated oil companies and independent producers of crude oil the ability to balance production with demand, at the administered price level of crude oil. The supply level was restricted to maintain artificially high and noncompetitive crude oil prices.[2]

3. The depletion allowance and other preferential tax treatment of crude oil profits made a dollar of pre-tax earnings yield more after tax profits than was the case for earnings from downstream pipelines, refining, and marketing operations.

The concentration of industry profits in crude oil activities following World War II led the petroleum companies to scramble to improve their crude oil self-sufficiency. By 1970 the five largest U.S. based international petroleum companies had increased their worldwide crude oil self-sufficiency to 109 percent from 95.8 percent in 1960. The next twenty-five largest integrated petroleum companies increased their crude oil self-sufficiency to 92.3 percent in 1970 from the 1960 level of 63.4 percent.[3]

This strategy of concentrating industry profits in first one and then another level of industry activity worked to the advantage of the integrated petroleum companies and to the detriment of the downstream independent refineries, terminal operators, and marketers of gasoline. For the integrated petroleum companies with a large degree of crude oil self-sufficiency, the high crude oil price simply resulted in downstream subsidization of refining and marketing activities from crude oil profits. To integrated refineries the price of crude oil was simply a transfer price. However, for independent refineries, the high, administered crude oil price was a real price that increased its raw-material costs. This placed the independent segment of the petroleum industry at a severe competitive disadvantage and permitted domination of the petroleum industry by a few companies. To the integrated petroleum companies, marketing, and to a lesser extent refining, became primarily a conduit for cashing in on highly profitable crude oil production. According to Mr. R. G. Follis, Chairman of the Board, "Standard of California's policy is and has been to use marketing to dispose of crude."[4] The orientation of marketing was primarily on volume and not upon return on investment.

The "profit game plan" of the integrated oil companies of maintaining artificially high crude oil prices and subsidizing their downstream operations faltered in the early 1970's. The international petroleum companies' control over crude oil was rapidly eroding and high profit was being squeezed from crude oil activities. The integrated oil companies' strength of position was no longer in crude oil, but in transportation, refining, and marketing activities, where control was solid and profit needed to be improved. Similarly, in the United States profits had to be made downstream since crude oil prices had been effectively frozen for several years, while exploration and development costs had risen significantly.

With these new conditions the logic of having crude oil carry the profit burden for the rest of the industry activities no longer made much sense. Crude oil profits were now harder to acquire, and it was increasingly difficult to link crude oil to unprofitable downstream activities. Also, the days of plentiful crude oil supply seemed to have passed. Unprofitable downstream operations were now hindering rather than furthering the sales of a company's own profitable crude oil. An example of this was Phillips Petroleum Company, which in the 1930's found no market for its crude oil in the United States and was forced to integrate downstream into refining and marketing. This situation reversed as integrated petroleum companies came to be able to sell every barrel of crude oil at a profit without having to subsidize unprofitable downstream operations. When Getty dumped its West Coast marketing operation in 1966 and Signal Oil and Gas did likewise in 1970, a significant reason given for both moves was a desire to concentrate on the production of crude oil.

The giant petroleum companies, however, have huge downstream investments which they cannot dispose of as easily as did Getty and Signal Oil. Executives of major petroleum companies such as Gulf, Arco, and Continental have recently underscored the need to put downstream investments on a sound, profitable footing. Furthermore, the major petroleum companies are committed to vertical integration, for they have grown and prospered within this strategy in the past. Time and again they have been able to use their financial muscle to push profits into one sector of the industry after another, and, in the process, they have been able to squeeze out their financially weaker competitors who cannot afford to develop balanced, vertical integration.

End of Oil Colonialism

As the 1970's opened, the stage was set for rapid loss of control by the international oil companies over crude oil production in the major oil producing and exporting countries. Several of the important factors leading to the loss of control and the erosion of earnings from international oil production are outlined below.

In September 1960 the major oil exporting countries formed the Organization of Petroleum Exporting Countries (OPEC) to bargain collectively with the oil companies. Through the 1960's the international oil companies maintained fairly good relations with the OPEC. As the 1970's opened, however, the relationship rapidly deteriorated. There were a number of significant setbacks over the four-year period from 1970 to 1973 that had serious, long-term consequences for the profitability of international production of oil by major petroleum companies.

Perhaps the single, most important incident was the military overthrow of the Libyan government by Colonel Muammar el-Qaddafi in 1969. Qaddafi rejected the level of production payments that his country was receiving for oil produced by the international petroleum companies in Libya. By 1970 he succeeded in a campaign that forced the oil companies to increase the tax reference price of crude oil exported from Libya. This was accomplished by pressure techniques involving the use of production cutbacks starting with one of the most vulnerable of the independent concessionaires, Occidental Petroleum Company. After Occidental yielded to the demand for increased prices and revenues per barrel, the other companies fell into line.

Libya's success in raising the payments it received on production triggered other Middle Eastern states to make similar demands for increased prices and production payments. After months of meetings, and an ultimatum from the Shah of Iran, the Tehran Agreement was reached on February 13, 1971, concerning Persian Gulf crude oil. This agreement called for annual price increases of five cents per barrel of oil plus a 2 percent annual inflationary factor. On April 2, a similar agreement was signed at Tripoli concerning Mediterranean crude oil.

By October of 1971 the OPEC countries were already dissatisfied with the price agreements signed at Tehran and Tripoli and were calling for higher crude oil prices as a result of devaluation of the U.S. dollar. On January 20, 1972, in Geneva, a supplemental agreement was signed to the Tehran Agreement which resulted in an increase in the posted price of crude oil of 8.49 percent and provided for future adjustments associated with changes in relative currency values.

In the fall of 1972 Libya kicked off a new OPEC campaign for participation in the ownership of the concessionaire's operations in OPEC countries. The oil companies officially met in Geneva on January 21 and 22 to begin initial discussion of such participation. Setting the stage for this meeting was Libya's expropriation in December 1971 of British Petroleum's 50 percent share of the Sarir field. This was followed on June 1, 1972, by the Iraqi government's nationalization of the northern Kirkuk field owned by Iraq Petroleum Company.

Sheikh Zaki Yamani, oil administrator for Saudi Arabia, was appointed as chief negotiator of the participation concept with the oil companies. These negotiations led to the "New York Agreement" of October 26, 1972, which resulted in the transfer of 25 percent ownership of oil companies' operations in the OPEC countries to the foreign nationals, effective January 1, 1973. The agreement further stated that by January 1, 1982, the oil producing countries could acquire a 51 percent controlling interest in the oil operations within their country.

Late in May 1973, Iran wiped out the nineteen year consortium agreement involving on-shore Iranian oil production and took over state control of the oil industry. The oil companies, in exchange for yielding control of on-shore production, received a twenty-year guarantee and exclusive rights to the production of most of the oil from that area. In essence, the oil companies remained as service contractors to the national Iranian Oil Company.

Qaddafi of Libya continued on the offense. On June 11, 1973, Libya nationalized the assets of Nelson Hunt and completed the takeover of the Sarir field. Libya and Iraq's nationalization of certain fields stood as a warning to the oil companies of the consequence if they resisted the desires of the oil producing countries for more immediate control of their oil industries than specified in the original participation agreement. Taking the lead once again, Libya announced on August 11, 1973, that it was nationalizing 51 percent of U.S. oil companies' operations. Kuwait confirmed that it also wanted an immediate 51 percent control of the oil industry within its territory. On October 6, at the outbreak of the October War, Iraq nationalized Mobil, Exxon, and Shell's interests in Basrah Petroleum Company from fields in southern Iraq. Finally, Saudi Arabia announced in mid-November that it was taking 51 percent control of the Arabian-American Oil Company (Aramco) and was not waiting until 1982, as had been agreed upon less than a year before.

Besides the rejection of the "New York" participation agreement and the demand and takeover of 51 percent of ownership, the OPEC countries also rejected the 1971 Tehran price agreements. In January 1972 the Geneva I pact, tying crude oil prices to fluctuations of major currencies to compensate the producing countries for two dollar-devaluations, was signed. In June 1973 signing of the Geneva II pact increased the posted price of crude oil by 11.9 percent. In September a new campaign was launched for still higher crude oil prices. The OPEC countries announced a 66 percent increase in the posted price of crude oil on October 16, 1973. This was rapidly followed by a doubling on the posted price by the six major Persian Gulf producers, effective January 1, 1974. The rapid increases in the posted price of Arabian oil over a three-year period can be observed in Table 4–1.

TABLE 4–1

Prices of Arabian Light Crude Oil

August 31, 1970	1.80
February 15, 1971	2.18
June 1, 1971	2.285
January 20, 1972	2.479
January 1, 1973	2.591
April 1, 1973	2.742
August 1, 1973	3.066
October 5, 1973	5.119
January 1, 1974	11.651

As a crowning blow, on October 17, 1973, eleven oil-producing countries announced a cutback in production of at least 5 percent per month and a total embargo on sales to the United States and Holland in retaliation for their support of the Israeli position in the Middle East crisis. The reduction of oil production and the embargo on oil sales were designed to force Israeli withdrawal from territories that it acquired from the 1967 and 1973 wars with the Arabs.

Deterioration of International Production Profit

The unfavorable set of developments confronting the international oil companies in the primary oil producing and exporting countries amounted to an ending to the long period of "oil colonialism." The oil producing and exporting countries decreed majority ownership and took over effective control of the oil industry within their respective boundaries late in 1973. This rise of oil nationalism has and will continue to have a long-run adverse impact on the Eastern Hemisphere earnings of the international oil companies.

The year 1972 was the worst profit year in a decade for the international oil companies' Eastern Hemisphere operations. Indicative of the problems of the international oil companies, the rate of return for three large European based international oil companies

—British Petroleum, Royal Dutch/Shell, and CFP—fell from 9.8 percent in 1971 to 4.9 percent in 1972.[5]

The earnings records of the seven sisters—Exxon, Texaco, Mobil, Gulf, Standard of California, Royal Dutch/Shell, and BP—reveal the downward pressure on profits from Eastern Hemisphere operations of the largest international oil companies. The turning point in the Eastern Hemisphere operations of the seven sisters can be observed from Figure 4–1 and Table 4–2, which show the falling earnings per barrel of oil produced and the increased production payments. Earnings per barrel at the start of the OPEC conflict in 1970 fell from 33.6 cents to 28.3 cents in 1972—down 5.3 cents. Conversely, production payments from 1970 through 1972 increased from $0.865 per barrel to $1.341—up $0.476. As this occurred, return on Eastern Hemisphere investment in 1972 fell to 10 percent (see Table 4–1), from an

FIGURE 4–1

Eastern Hemisphere Earnings and Product Payments

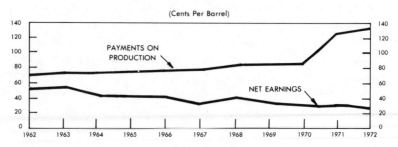

Source: First National City Bank.

average of 12.2 percent for the four prior years. The low 10 percent return from Eastern Hemisphere oil operations can also be contrasted to the 11.1 percent for all manufacturing in the United States in 1972 (see Figure 4–2).

Over the ten-year period from 1962 through 1972, the rate of return of Eastern Hemisphere operations of the seven sisters fell from the peak level of 14.1 percent in 1963 to 10 percent in 1973. Paralleling the decline in return on investment, earnings per barrel of oil produced fell from 56.3 cents in 1963 to 28.3 cents in 1972. While earnings per barrel were approximately halved from 1963 to 1972, the reverse situation was true for production payments to the government, which almost doubled over the ten years, from 70.9 cents

TABLE 4-2

Eastern Hemisphere Operations of Seven Sisters[a]

	1962	1963	1964	1965	1966	1967	1968	1969	1970	1971	1972
Gross production (million barrels)	2,310	2,540	2,882	3,235	3,627	3,919	4,386	5,016	5,711	6,563	7,058
Product sales (million barrels)	2,051	2,235	2,473	2,722	2,937	3,210	3,519	3,834	4,311	4,403	4,656
Net earnings ($ millions)	1,227	1,429	1,245	1,353	1,490	1,451	1,782	1,818	1,917	2,236	1,995
Net worth on Jan. 1 ($ millions)	9,353	10,115	11,232	12,027	12,929	13,542	14,245	15,446	16,249	17,370	19,882
Return on net worth (%)	13.1	14.1	11.1	11.2	11.5	10.7	12.5	11.8	11.8	12.9	10.0
Earnings per barrel (cents)	53.1	56.3	43.2	41.8	41.1	37.0	40.6	36.2	33.6	34.1	28.3
Production payments to govts. ($ millions)	1,637	1,908	2,167	2,471	2,798	3,131	3,656	4,209	4,938	8,296	9,466
Payments per barrel (cents)	70.9	75.1	75.2	76.4	77.1	79.9	83.4	83.9	86.5	126.4	134.1

Source: First National City Bank.

a. BP, Exxon, Gulf, Mobil, Royal Dutch/Shell, Standard of California, and Texaco.

FIGURE 4–2

Comparison of Percentage Return on Investment

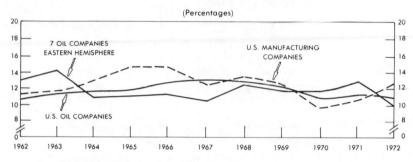

(Percentages)

Source: First National City Bank.

in 1962 to 134.1 cents per barrel in 1972. The production payments have since spiraled, increasing to seven dollars per barrel of oil beginning January 1, 1974.

The days of highly profitable foreign crude oil production by the international oil companies from the primary oil producing and exporting countries are likely gone forever. These countries have effectively confiscated 51 percent or more of the ownership of oil company assets. OPEC will now take a majority of the profits and in some cases all the profit where 100 percent nationalization has occurred.

If the international oil companies could increase the prices of crude oil enough to make up for the loss of ownership, they could restore much of the lost earning potential. This appears to be unlikely over the long run since the producing countries are endeavoring to control the profit per barrel of crude oil. This was one of the purposes of the 66 percent increase in the posted price of crude oil decreed on October 16, 1973. The OPEC countries intended to increase the production payments and reduce the earnings of the oil companies. Intelligence work by the OPEC countries indicated that more than 80 percent of the oil companies' profit at that time was being earned from production, and they wanted to change that situation. At the other end, the consuming countries are increasing their controls over the oil countries to impair their ability to pass through price increases.[6]

In summary, the long-run effect of the loss of 51 percent ownership over production assets in several Middle East countries and the limitation on per barrel oil profit will be further pressure on return on investment from Eastern Hemisphere oil production of the seven sisters. With return on investment in major oil producing and exporting countries falling below the competitive rate in recent years, and with long-run projections for a continuing squeeze, the international oil companies are no longer in a position to subsidize downstream investment in refining, distributing, and marketing from foreign-produced oil.

Some idea of the extent of past downstream subsidization with production profit can be obtained from an *Oil & Gas Journal* article entitled "Majors Earn Less from E. Hemisphere," dated October 26, 1970 (p. 41). In discussing "Downstream Depression" of refining and marketing in the Eastern Hemisphere, the Journal stated that, "In Western Europe, an average return on net worth of only 4 percent in 1968 fell to a mere 1.6 percent in 1969. The high had been 5.6 percent in 1963." Similarly, an article in *Business Week* pointed out how fat crude oil profits used to carry downstream refining and marketing operations.

> "For the first 50 or 60 years of our existence," recalls Dorsey [Chairman of the Board of Gulf Oil Company], "the oil business was uncomplicated." With a 50 percent ownership in vast oil reserves in Kuwait, fat profits were made in the 1950's and 1960's from low-cost crude production. It did not really matter much whether Gulf's refining and marketing operations made money. They were just an outlet for the crude.[7]

The *Business Week* article went on to explain that all of this has rapidly changed.

> The last few years have revolutioned the oil business. There is no longer free and easy access to Mideast oil. . . . The oil rich nations now have the upper hand, nationalizing or demanding equity participation in oil companies, clamping limits on oil exports, and pressing for higher prices for crude.

Oil Production Doldrums in the United States

The domestic oil industry stagnated during the 1960's and early 1970's. One of the reasons for its depressed state was the discovery of vast quantities of low-cost crude oil in the Middle East. The much higher cost of domestic production placed the U.S. industry at a severe competitive disadvantage.

The Mandatory Oil Import Program, a quota system to restrict oil imports, was established by Presidential Proclamation on March 10, 1959. The technical ground for the quota system was that national defense interest required the maintenance of a strong domestic oil industry, and administrators of the program were responsible for determining whether or not the price increases could be justified on national security grounds. The Mandatory Oil Import Program protected domestic oil production from the price-depressing competition of low-cost imports. However, it also placed an effective upward ceiling on the domestic price of crude oil.

During the fourteen-year life of the oil import quota system, the price of domestic crude oil was relatively constant, as can be observed from Figure 4–3. From March 1959 through April 1973, the domestic

FIGURE 4–3
Cost of Domestic Crude Oil

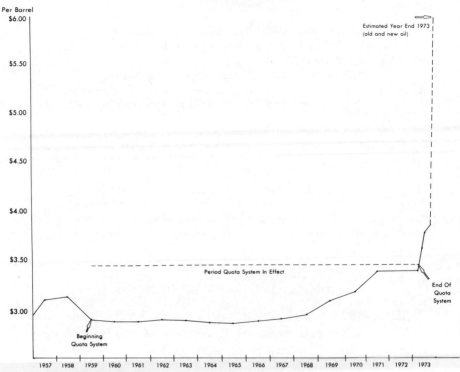

Source: Independent Petroleum Association of America—annual data, 1957–1972, and monthly for 1973.

price of crude oil increased from $2.90 per barrel to only $3.40 per barrel. For the ten years from 1959 to 1968, the price of domestic crude oil had been maintained at $2.90 per barrel. Early in 1969 there was a modest increase of fifteen cents per barrel in the price of domestic crude oil and late in 1970 the domestic price of crude oil was increased by twenty-five cents a barrel. However, much of this second increase was to compensate for the reduction of tax benefits. In 1970, the depletion allowance was reduced from 27.5 percent to 22 percent, a minimum tax was established, and the investment tax credit was cut. Furthermore, wage and price controls were implemented in the United States on August 15, 1971, and the price of domestic crude oil remained frozen through Phases I, II, and III. It is clear from the record that, while the quota system protected the domestic price level, it did place an effective lid on crude oil price increases.

Effects on Oil Production Profit

Although crude oil prices remained almost constant over the fourteen-year period of the oil import quota system, the costs of exploration and production were increasing. This cost-price squeeze was eating into highly profitable crude oil operations. Toward the end of the fourteen-year period, exploration costs were accelerating and the price received for new oil was said in many cases not to be covering the replacement costs.

The American Petroleum Institute (API), the primary lobby for the major petroleum companies, in October 1972 summarized the price-cost squeeze on oil production for the Interior and Insular Affairs Committee of the Senate during its investigation of the declining trends in exploration for oil and gas as follows:

> Several factors have raised the cost structure of the oil producing industry in recent years, but market prices have not risen sufficiently to compensate for the additional burden.[8]

One of the fundamental reasons for the increasing cost of new onshore oil is the age of the oil industry in the United States. Many of the larger, more readily accessible fields were long ago discovered and more recent exploratory drilling has yielded fewer and less productive oil fields. Deeper drilling has been necessary and has increased the cost of exploration while inflationary pressures have driven up drilling costs.

The cost of existing production in many of the older fields has also been increasing. As field production has declined the cost per barrel of producing oil has increased. To slow the rate of decline, costly secondary methods of production have been employed. This decreasing production and increasing cost has been eating into the profitability of old domestic oil.

The decreasing opportunities for onshore production from the Continental United States have led to the search for oil in less accessible offshore locations and in less clement areas such as Alaska. The special drilling equipment and expenses required for these efforts have driven up the cost of drilling several times over that of onshore wells in the lower forty-eight states. Such onshore wells had an average cost of around $80,000 in 1971. In comparison, offshore wells cost $530,000 and Alaskan wells $1,260,000.[9] In addition, because the areas in which this oil is discovered are less accessible, transportation costs in moving the oil to the refineries where it is needed have grown.

Environmental problems have also greatly increased the cost of exploration and production from offshore areas and in Alaska. A moratorium was placed on drilling in the Santa Barbara channel in 1969 following the great pollution damage which resulted from the blowout and loss of control over an offshore well. In January 1968 one of the largest oil strikes in North American history was made by Atlantic Richfield near Prudhoe Bay on the North Slope of Alaska. Although billions of dollars have been spent to acquire leases and to drill wells, this oil will not reach the market before 1976 because of environmental litigation which had postponed construction on the Alaskan pipeline until 1974.

Adding greatly to the already costly process of offshore and Alaskan exploration has been the "bonus bidding" method of leasing these federally controlled lands for oil and gas production. In the late 1960's and early 1970's, astronomical sums were spent on bonus bids. The record $1.6 billion Louisiana lease sale in December 1972 was part of a total $2.25 billion spent on Louisiana Gulf Coast leases for the year. The June 1973 lease of offshore Texas acreage came within $100,000 of the record Louisiana sale.

The intense pressure on petroleum companies to increase their reserves to stay in the petroleum business has been frustrated by the Nixon administration's policy of making small and infrequent lease sales. This policy has led to intense competition for drilling acreage and to the huge prices paid for hunting licenses to explore for oil and gas. The bonus paid per acre would have been much lower had greater acreage been leased through larger and more frequent sales. The government could have easily done this as less than 3 percent of the offshore land has been leased for exploration.

The parceling out of small acreage for lease sale by the Nixon administration has also had a number of undesirable side effects. While maximizing the short-run cash flow to the U.S. Treasury, it has slowed the exploration and production for new oil and gas. The $1.6 billion spent for the record bonus bid in December 1972 would have paid for 21,330 oil wells or 11,405 gas wells based upon average 1971 costs.[10] Furthermore, huge bonus bids work to the advantage of those that already have lower-cost existing oil, and so it stymies new competition in the industry. Over the long run, only those companies which can obtain windfall profits on large existing reserves can shoulder the large financial burden of the bonus bidding procedures. Thus it can be argued that the present bonus bidding system is not managed in society's best interest as it retards both the rapid development of new crude oil and the channeling of that crude to the more efficient but less wealthy competitors.

The effect of the administration's bonus bidding system has been to further drive up the cost of exploration, making the production of new oil extremely expensive. This policy, combined with the lid on crude oil prices through the first quarter of 1973, forced the major petroleum companies to look downstream for profits in refining and marketing for the first time in many years.

Downstream Profits

The declining profitability of crude oil operations was not the only reason for reversing the long-standing policy of subsidizing downstream operations. Capital needs were spiraling and the Chase Manhattan Bank estimated that one trillion dollars would have to be spent in the non-Communist world petroleum industry from 1970 through 1985. This was twice the magnitude of expenditure of the preceeding fifteen years. To reach this level of spending, energy economists predict that $600 million will have to be internally generated and $400 billion obtained from the capital markets. With the poor economic condition of the petroleum industry at the beginning of 1973, it was unlikely that either of these figures would be achieved because of the competition for capital from other, more profitable segments of the world economy.[11]

The petroleum companies' earnings from downstream activities has been a closely guarded secret. Most requests for such information go unheeded. For example, the Interior and Insular Affairs Committee of the U.S. Senate investigating the declining trends in exploration for oil and gas requested information from the API on rate of return from production and manufacturing. The only information presented has been on total return from integrated petroleum company operations—nothing permitting a comparison of return on investment from production and downstream operations. In 1972 the Federal Trade Commission tried to subpoena information on rates of return from production and downstream investment but was also unsuccessful. Efforts by the Senate Commerce Committee and Senate Permanent Investigating Subcommittee to obtain this information in 1973 also failed.

The integrated petroleum companies have good reason to be so sensitive about their earnings from production in comparison to downstream investment and to keep such information from government and public scrutiny. Serious antitrust questions would be raised if it were shown that crude oil prices had been administered at artificially high levels and had been used to subsidize downstream operations in competition with independent refineries, terminal operators, and marketers. Furthermore, public disclosure of the relatively high rate of return from production would make it difficult to continue to administer artificially high crude oil prices. It would also result in increased pressure from Wall Street on the petroleum companies to increase downstream prices and revenue in areas where competitive pressures are much greater than those in production activities.

Even without access to more complete return on investment data of the giant integrated petroleum companies, it is possible to show on an indirect, piecemeal basis that crude oil profits have been historically used to subsidize downstream operations. For the international operations of the integrated petroleum companies, downstream subsidization can be inferred by comparing the return on investment of the overall integrated companies' operations with that of thirty-six companies principally involved in transportation, marketing, and refining. The return on investment of twenty-five refining and marketing companies in Western Europe over a ten-year period from 1963 to 1972 has ranged from 1 to 6 percent. The profitability of six Australian companies has been from 4 to 9 percent, and five Japanese companies, with a high debt-to-equity ratio, reported a return of 11 to 16 percent. The overall rates of return "for all thirty-six *downstream* companies have remained extremely low—generally 3 to 5 percent."[12] The integrated oil companies, however, have maintained approximately a 12 percent return on investment. It is evident that the international oil companies have subsidized downstream transportation, refining, and marketing activities and in the process have exerted extreme pressure on their independent competitors, preventing them from earning a competitive rate of return.

An accounting employee of Exxon-Europe explained that for ten years prior to 1972, corporate pricing strategy had been devised to take profit in tax-sheltered Eastern Hemisphere crude oil operations and to operate refining and marketing on close to a break-even basis. Data studied on Mobil's joint venture with Exxon in Iran during the early 1960's showed that Mobil earned a rate of return from oil production after all taxes in excess of 30 percent on investment—twice the companies' overall average.

The past practice of earning little or no profits from European refining and marketing with big crude oil profits was also explained in a *Forbes* magazine article in early 1972, part of which is reproduced below.

> The international oil companies are all refiners and marketers as well as producers, but historically the profits have always come from producing. Until 1968, profits from refining and marketing amounted to zero or less in Western Europe. . . . The companies have resorted to a variety of devices . . . to keep profits in refining and marketing to a minimum in order to keep taxes to a minimum.[13]

Similarly, a *Business Week* article in late 1973 concerning Gulf Oil Company problems discussed the big crude oil profits of the past and the lean earnings from downstream operations. The article stated:

> . . . fat profits were made in the 1950's and 1960's from low-cost [foreign] crude production. It didn't really matter much whether Gulf's

refining and marketing operations made money. They were just an outlet for the crude.

But the last few years have revolutionized the oil business.[14]

A Continental Oil Company executive stated in an affidavit in the 1970 Tidewater/Phillips merger suit by the Justice Department that Douglas Oil Company had been purchased in 1960 as an outlet for foreign crude oil. Facts showed that while it had been operated for ten years on close to a break-even basis, the changing economics of the petroleum industry forced Continental to dispose of its unprofitable operations and to restructure downstream operations to make them more profitable.

A director's budget for Cities Service Company in the 1960's showed that marketing lost millions of dollars per year and was being carried by crude oil profits. Data analysis for Tenneco showed that refining and marketing for a five-year period averaged out as a loss operation. From its combined refining, distribution, and marketing properties, Phillips Petroleum Company earned only 1.98 percent in 1969, .69 percent in 1970, .17 percent in 1971, and 1.92 percent in 1972.

The return on investment of Skelly Petroleum for exploration and production in comparison to refining and marketing for the years 1970 through 1972 is shown below:

Return on Investment
(in percent)

Year	Exploration & Production	Refining & Marketing
1970	11.4	1.7
1971	11.3	1.3
1972	8.6	2.2
Average	10.4	1.7

The Skelly financial data clearly shows the subsidization of downstream operations with crude oil profit. It is unfortunate that similar divisionalized data could not be obtained for all petroleum companies and be made public information.

The deterioration of large production profits, both domestically and internationally, was destroying the integrated oil companies' "profit game plan" of concentrating industry profit in less competitive, tax sheltered, crude oil activities. With eroding crude oil profit and accelerating capital needs it was no longer feasible to continue downstream subsidization. To improve overall corporate profitability which had been deteriorating from 1968 through 1972, it had become necessary to put downstream refining, distribution, and marketing operations on a more profitable basis.

At the beginning of 1973 editorials in the major petroleum trade journals pointed to the need to breathe profit into downstream operations. On January 1, 1973, an editorial in the *Oil & Gas Journal* stated that changing conditions in the petroleum industry

> . . . call for a new set of priorities creating stable profit centers in re-fining and marketing. The time has long gone when a company can underwrite refining—marketing losses with production profits.[15]

Frank Breese, editor of the *National Petroleum News (NPN)* stated in the January 1973 issue of that journal that:

> there is more pressure than ever on return on investment . . . market-ing must make a bigger contribution to company earnings; and changes in the coming 12 months will be greater than those of the recent past.[16]

A year earlier, Keith Fanshier, editor of the *Oil Daily,* under-scored the need to restructure petroleum industry profits and make downstream activities profitable. In his editorial entitled, "The Look Downstream," he said:

> If the petroleum industry is to become more fully rewarded and finan-cially appreciated, it would seem such condition would have to come about . . . by more nearly adequate recompense from its downstream operations.
>
> . . . More, and more, authoritative observers are coming to believe, and to express the view, that a truly economic future for the industry lies in its ability to enhance its downstream returns.
>
> A realistic view indicates that truly prosperous downstream re-sults have seldom existed, except for relatively short intervals. Yet, this fact should not be judged to foreclose the possibility of its being brought about. . . .
>
> . . . any profit adhering to the marketing function particularly, and to some extent, the manufacturing are minor.
>
> Downstream "profit centers" envisioned, created and enforced will have to be a decisive key to the industry's prosperity of the future. This will require a high order of strategic, tactical and operating deter-

mination, skilled salesmanship and protection of a really profitable refined-product-price structure. This will help the crude oil market, as a firm crude market, in turn, can help it.[17]

Major oil company executives also emphasized that new conditions required that downstream refining and marketing pay its own way. Robert B. Phillips, executive vice-president of Gulf Oil Company—U.S., was quoted in *Business Week* as saying, "For a long time, we had an excess of crude oil and the name of the game was to move as much crude oil as possible through our marketing system."[18]

The article went on to point out that the declining profitability of crude operations was forcing Gulf to place downstream investments on a paying basis. One Gulf executive was quoted as saying, "We intend to make our marketing investments stand on their own two feet."[19]

The article further stated that the "same message is being carried overseas" as a consequence of declining profitability as well as uncertainty over future prospects.

Mr. W. H. Burnap, executive vice-president, and Mr. Thomas Sigler, vice-president of Continental Oil Company, told the Cost of Living Council in early 1973 that the petroleum industry has realized that it must now make profits from refining and marketing without relying on profits from crude oil.[20] Similarly, Robert T. McCowan, senior vice-president of Ashland Oil Company, stated that:

> The old adage, "exploration and production make the profits while refining and marketing break even," can no longer be valid. . . .
>
> . . . gasoline marketing is beginning to be looked upon as a part of the petroleum industry which must, on its own, make profit contributions relative to the investment involved without being subsidized by other segments of the business. No longer can gasoline retail markets be considered just a means of maintaining high crude oil sales and refining runs.[21]

Measures Designed to Improve Downstream Profitability

The integrated petroleum companies experimented with a number of programs designed to improve their return from downstream investment. These programs included:

1. Entry into private brand, discount gasoline marketing through the use of "secondary brands";

2. Withdrawal from particularly unprofitable marketing areas; and,
3. Elimination of costly price allowance programs which permitted major brand jobbers and dealers to lower their prices.

The reduction of price allowance, elimination of high loss operations, and the decrease of operating costs through secondaries were all measures that some companies hoped would help improve their downstream return on investment.

Major Secondary Brands

One of the purposes of the majors' secondary brand operations was to try to stop losses in some of the weaker, less profitable major-brand territories and outlets. The hope was that these unsuccessful major brand outlets might survive as unbranded, higher volume, price discount operations. Another important reason for the movement of integrated companies into secondary brands was the continuing volume growth of lower priced, private brand gasoline. If discounting could not be stopped, then the majors had no choice but to become involved if they intended to stay integrated through marketing. Other reasons for "secondaries" included keeping closed major brand stations from being reopened by independents as private brand discount outlets and the use of secondaries as fighting brands to combat the growth of high volume, profitable, private branders.

The major oil companies have typically converted existing outlets into secondary brand stations. In the conversion process the major brand sign has been replaced with one of a secondary brand, the station has been repainted, trading stamps and company credit cards have normally been dropped, the service bays have usually closed, and the TBA (tires, batteries, and accessories) line has been dropped. On occasion new secondary brand operations would be opened to determine the response that a first class, major run, secondary brand operation might receive.

Most of the major integrated oil companies have experimented with secondary brands. A list of several of these secondary brand operations by name is presented in Table 4–3. The purpose of each name has been to make a clear-cut differentiation between the discount, price oriented, private brand method of marketing and the high priced, heavily promoted, major brand approach.

Phillips had almost nine hundred secondary brand stations in

TABLE 4–3

Secondary Brands of Integrated Oil Companies

Integrated Oil Company	Secondary Brand
Exxon	Alert
Texaco	Satellite
Gulf	Bulko, Easy Go, Economy, Transport
Mobil	Sello
Standard of California	PDQ
Shell	Ride
American Oil Company	Whale
Arco	Award, Rio Grande, Rocket, Wesco
British Petroleum	Gas 'N' Go
Phillips	Bingo, Black Jack, Blue Goose, Community, Excel, Finest, Jack Por, Jiffy, Pathfinder, Seaside
Sun Oil Company	Freeway, Premier, Quality Oil
Conoco	Douglas, Equal, Fast Gas, Gas-O-Less, Kay O, Western
Skelly	Surfco

April 1973 and was one of the largest oil companies with expanding secondary brands. During the 1950's Phillips put a big effort into expanding its market territories and became one of the few fifty-state gasoline marketers. While Phillips had the distinction of being a fifty-state operation, its marketing operation was relatively weak. In many instances Phillips simply did not have the high quality stations, the appropriate locations, or the market representation to operate on the quality major brand marketing plane. This was particularly true with Phillips' West Coast operation, most of which had been purchased from Tidewater in the mid-1960's. Generally, the Tidewater outlets and locations were of relatively poor quality and Phillips' West Coast marketing simply could not compete with the quality of Shell, Socal, Mobil, Arco, and others.

Mobil's secondary "Sello" brand experiment was another attempt to improve the profitability of stations not making it as major brand outlets. As low-price discount operations, the Sello outlets developed much higher volumes than they had as Mobil brand stations. Similarly, American converted several of its newer ranch style stations that were not generating acceptable profit in Louisville, Kentucky, to its secondary Whale brand.

Major brand marketers also hoped that secondary brand outlets might help with the problem they were experiencing of closed and abandoned stations being reopened by independents as low-price, private brand outlets. By operating the stations as secondary brands, the major brand oil companies could at least control prices and gallonage. For example, Herb Timms, an independent private brand marketer, reopened about two hundred old major brand stations as Petrol Stop and Petrol Express outlets over a five-year period. Other independents were doing the same thing, and these reopened stations were becoming a thorn in the side of major brand marketing.

Another purpose of major secondaries was as "fighting brands," intended to drain business and profits from lucrative, high-volume private brand stations. In the East Exxon frequently built Alert outlets near the high volume, high profit Hess stations. In Dallas Mobil opened several of its secondary Sello brand outlets near successful private brand outlets.

Withdrawal from Marketing Areas

Another major oil company approach to improving downstream refining and marketing profitability was to withdraw from areas of poor marketing performance. This reversed the market territory expansion trend of the 1960's, when several companies exerted considerable effort to increase their market coverage even to the extent of selling their brand in all fifty states. Some of the major oil company withdrawls are described below.

Arco. The Justice Department required that certain overlapping marketing properties resulting from the 1969 Atlantic-Richfield–Sinclair merger be sold. This sale left Arco with relatively thin market representation and a very unprofitable marketing operation in the Southeast. In mid-1971, Arco announced its decision to withdraw from major areas in the Southeast and South. Discussing the pullout, Arco marketing vice-president Charles Walsh stated, "total sales of gasoline amounts to less than 3 percent of the gasoline volume in each of the states involved,"[22] and he continued, "experience indicates that below 7 to 8 percent [market share], both supplier and retailer find it . . . difficult to generate and sustain a reasonable profit."[23] Arco president T. F. Bradshaw further commented on the withdrawal by stating, "We plan to operate only in areas where present or potential market share indicates a proper return on investment."[24]

BP. British Petroleum entered the United States in early 1969 with the purchase from Arco of two refineries and 10,000 Sinclair stations in sixteen states. In mid-1969, BP (USA) announced plans to merge with Sohio. Faced with thin market representation and bad pricing problems in the Southeast, BP—Sohio decided, as had Arco, to withdraw from the branded market in North and South Carolina. In early 1973, BP–Sohio further announced that it was withdrawing from the Southeast and selling more than a thousand stations to American Petrofina.

American Oil Company. American entered the West Coast in the early 1960's, along with Gulf, Exxon, and Continental. Frequent price wars and thin market representation made this a very unprofitable marketing venture for American. In March 1972 American announced that it was selling a group of stations to Shell in California, Oregon, and Washington, as part of a 300-station cutback on the West Coast. The remaining stations were sold in groups to major brand and independent marketers.

Phillips Petroleum Company. On May 22, 1972, G. J. Morrison, marketing vice-president for Phillips Petroleum, stated that "we are constantly studying our trading area . . . where we have a low percentage of the market. We know that if we cannot expect to eventually have a reasonable percentage of the market we should seriously consider leaving. . . ."[25] By mid-1972, Phillips management had examined its northeastern market and decided that it lacked adequate regional gasoline supplies, that its market share was too low, and that the company should withdraw from the market. The fourteen-month withdrawal affected 1,400 stations and 300 million gallons of gasoline. Phillips president, W. F. Martin, blamed the withdrawal on "the unprofitable experience our company has had in this area as a supplier."[26]

Gulf. In October 1972 the board of directors of Gulf Oil Company approved plans to sell 3,500 retail outlets and related distribution facilities in fourteen midwest and six northwestern states. In its 1972 annual report, Gulf stated that these properties were "unprofitable or marginal investments which were not achieving a satisfactory rate of return."[27] The reasons for this poor rate of return were intense competition with private branders and poor market representation in the area.

Other. Conoco announced in late 1972 that it was withdrawing from all branded marketing in East Texas. Executive vice-president William Burnap explained his company's move by stating that the gasoline industry "has realized it must make profits now on refining and marketing and [not] just . . . on crude oil."[28] In early 1973, Sun Oil Company announced plans to withdraw from three hundred outlets and related distribution facilities in seven midwestern states and Tennessee. Sun's marketing vice-president, T. A. Burtis, explained that the withdrawal occurred "because the earnings . . . realized in these areas have been insufficient to justify the continued commitment of capital."[29]

Eliminate Price War Subsidies

The marketplace was racked by gasoline price wars during much of the period from 1969 to 1972. This price warfare was taking a heavy toll on corporate profits. For example, by late in 1972, Los Angeles had suffered through 270 days of price wars and East Texas prices were only "normal" one week in four. Overall, it was estimated that during 1971 and the first half of 1972, price wars were costing the oil companies close to $2 billion on an annual basis.

The price wars were being fed by a "price allowance" system, whereby the major oil company suppliers reduced prices to their dealers, permitting the dealer to cut their prices at the service station. The way the price allowance system worked was to have the dealer absorb .3¢ and the oil company supplier .7¢ per gallon for every 1¢ reduction in the suggested retail price.

The major oil companies increased the retail price of gasoline on a nationwide basis twice during 1970. The increase was needed to offset rising marketing costs and to improve downstream profitability. However, with these increases the tempo of price wars rapidly accelerated and the overall profits of the oil companies actually deteriorated. One of the important reasons the price war problem accelerated was the loss of volume from the majors to the independents as the major brand price level rose. Huge price allowances had to be given to major brand dealers to keep the high cost, major-brand dealer system from being destroyed by the low priced, private-brand discount gasoline marketers. Price problems were further aggravated by the price cutting of certain major brand dealers—known in major brand circles as "mavericks"—who cut price against the norms of major brand behavior.

The enormous costs of the price allowance led some smaller integrated oil companies to introduce programs designed to cut the price subsidies and to improve profitability. Along this line, Mr. Burnap, executive vice-president of Continental Oil Company, stated that price allowances were "a key source of current marketing problems" and that "they prevented emphasis on the marketing approaches to public desires by preventing failure of inefficient ones."[30] In mid-December 1970 Conoco and Phillips lowered and eliminated price allowances in Idaho, Utah, Nevada, and parts of surrounding states in an experiment which lasted less than four months. Tom Sigler, Conoco's vice-president of marketing, stated that "other major oil companies apparently are not presently inclined to discard archaic pricing systems."[31] He went on to state that Conoco was not giving up its "search for improved pricing practices." Phillips, which experimented at about the same time with a new pricing plan designed to reduce price subsidies, was also forced to abandon its experiment. Other, smaller petroleum companies joined in the experiments with reformed pricing programs to reduce the costly price allowances.

On September 1, 1971, Continental started a new experiment in the price-torn Rio Grande Valley of Texas aimed at eliminating the costly price allowance system of major brand marketing. The experiment was a "three-tiered" pricing plan which included an unbranded terminal price for gasoline, a brand fee, and a cost for leasing service station facilities. This was followed in December 1971 by a new pricing plan introduced by Phillips in parts of Colorado and Wyoming. The new plan greatly reduced price allowances but did not totally eliminate them. In late January 1972, Conoco introduced what it called a "realistic" pricing program in Colorado, Missouri, and parts of Kansas. On February 1, 1972, Citgo announced its new pricing plan, which contained aspects of both Conoco's "Valley" terminal pricing program and Phillips limited price allowance formula. On December 1, the Citgo plan was extended into the upper Midwest operation. Other major oil companies, including American, Mobil, and Gulf, conducted limited tests of reformed pricing programs in late 1971 and 1972.

Problems in Improving Downstream Profitability

The major oil companies' efforts at improving the distressing profitability of downstream operations did not succeed. As can be observed from Figure 4–4, return on investment continued to decline from its peak 1967–68 level through 1972. The rapid upturn in profit-

FIGURE 4–4

*Return on Investment and Fluctuations in Wholesale
Price of Gasoline from 1960 to 1973*

A. Return on Investment of Major Petroleum Companies[a]

B. Monthly Average Wholesale Price for Regular Grade Gasoline for
   55 Markets[b]

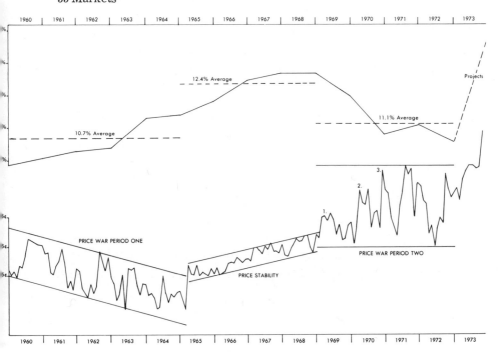

a. *Source:* The First National City Bank of New York.
b. *Source:* Platt's Oilgram Price Services.

ability in 1973 was largely a result of the petroleum product short-ages, not a consequence of the measures discussed.

It is doubtful that the major secondary brands did anything to improve the overall petroleum company profits and may even have had a negative effect. While it is true that the return from assets converted to secondaries often improved, the secondaries frequently aggravated price problems and probably contributed to worsening overall petroleum company profit.

The withdrawals were a step in the right direction toward im-proving oil company profits. They normally resulted in petroleum companies drawing back from marketing areas where losses were large and their future did not look good. The greatest shortcoming of this situation was that stations simply changed brands instead of going out of business. This did nothing to reduce the grossly overbuilt service station population.

Efforts to eliminate or reduce the "price allowances" which were feeding price wars were largely ineffective before the occurrence of the petroleum product shortages. Such giants of the industry as Exxon, Texaco, Mobil, Gulf, Socal, Shell, and American did not generally support a reform of the old price allowance system under which major brand marketing had grown up. Without their support, the smaller petroleum companies and independent refineries could accomplish relatively little toward eliminating subsidies and reduc-ing the frequency of costly price wars. For example, Tom Sigler of Continental said, "We are not so naive as to believe that this system [Conoco's experiment of operating without price allowances] can survive if other suppliers continue the practice of granting sizeable temporary price supports."[32] Both Continental and Phillips after a few months backed off on several of their experiments when the larger petroleum companies failed to support their initiative to eliminate price protection. Continental and Phillips jobbers and dealers were being hurt by competitors who were able to sell gaso-line at lower prices because of continued price allowances from their suppliers.

Price Wars

It may seem paradoxical that the larger petroleum companies would not go along with efforts to eliminate the price allowances and end the price wars. These price wars were cutting into profits and depressing return on investment. Yet the large integrated petroleum companies were "over a barrel" and basically unable to come to grips with "price allowance" and the price war problem.

Gasoline is the largest volume and most profitable product refined from crude oil. It is not surprising, therefore, to find the return on investment of the integrated petroleum companies being roughly correlated with relative stability of wholesale gasoline prices. Since most changes in branded dealer prices are underwritten by lowering the wholesale prices through granting dealer price allowances, changes in wholesale and retail prices closely parallel one another.

Over the fourteen-year period from 1960 to 1973, there were two major price war periods during which wholesale and retail gasoline prices fell sharply and frequently. As can be observed from Figure 4–4, one price war period was from 1960 to 1964. The price instability was often touched off by battles between the private brand outlets of the independents and major brand stations over price differentials. While most private branders wanted to continue to sell regular grade gasoline for two cents per gallon less than the major brand price, some major oil companies wanted to reduce the price differential to one cent or none at all. The second price war period, during which wholesale and retail prices also fluctuated violently, lasted approximately four years, from 1969 through 1972. This price war period was associated with the rapid growth of private brand, discount gasoline marketing and particularly with self-service marketing.

The return on investment of the major petroleum companies was depressed during both of these price war periods because of the billions of dollars they had to pay in "price allowances" to finance the price wars. As can be observed from Figure 4–4, return on investment averaged 10.7 percent for the first price war period and it averaged 11.1 percent for the second. The in-between period from 1965 to 1968 was one of infrequent price wars and relative price stability, and the rate of return on investment of the petroleum companies was much higher at 12.4 percent. Similarly, the cessation of price wars and the climb in wholesale prices which have resulted from the current petroleum product shortage have also resulted in a record return on investment.

An Archaic Marketing System

The primary problem of the majors' marketing system today is that it has been built up over the years by crude oil profit subsidies. In other words, marketing investment was never justified on an economic basis by the test of the marketplace and the buffering of competition.

As giants of the petroleum industry developed their marketing operations on a nonprice competitive basis, they greatly overexpanded their marketing systems in order to cash in their monopoly oil profits. Consider the statement in 1960 of D. Woodson Ramsey, vice president of marketing of Humble—a division of Standard Oil of New Jersey, concerning the excessive number of stations:

> The industry has more damn gasoline stations than we have customers —the oil industry problem is to find some way to deliver gasoline to the consumer other than by having service stations on every corner. It is too expensive.[33]

Ten years later, in 1970, the chairman of the board of Exxon, Ken Jamieson, made a similar observation when discussing the service station problem: "We are too standardized on every street corner, with the same structure and same merchandise. That isn't the answer."[34]

Also in 1970, Tom Sigler was talking about the "marketing overkill" and said:

> This attempt to push crude through service stations has been at least partially responsible for one of our greatest problems—that of too many service stations.
>
> In my opinion, the conventional service station representation in many parts of the country is much greater than is needed to serve the consuming public. It's a case of marketing overkill.[35]

Finally, Gulf Oil Chairman B. R. Dorsey stated before the New York Security Analysts that "the people would be better served by 125,000 stations," rather than the 250,000 stations he estimated exist.[36]

The major oil companies could see what was happening to their crude oil profits and saw the need to bolster marketing prices to improve return on investment. In fact, the petroleum companies increased the wholesale price of gasoline three times and by more than two cents per gallon during 1969 and 1970, as can be observed from Figure 4–4. This was more than twice the wholesale price increase and it took place in half the time of the gradual price rise from 1965 through 1968. It should be noted, however, that the three sharp price increases during 1969 and 1970 did not hold, and, following each price increase, the markets experienced chronic price wars. By mid-1972 wholesale prices had fallen across broad marketing areas to the 1967 level, at which time costs were much lower.

The difficulties experienced by petroleum companies in trying to lift their market prices were caused by their terribly overbuilt, inefficient, costly marketing system that could not stand the test of the marketplace. As the majors tried to increase their prices, they succeeded only in losing more and more of their market share. The ten largest petroleum companies dropped from an estimated 66.1 percent market share in 1968, to 59.5 percent in 1972, as shown in Table 4–4.

TABLE 4–4

Declining Market Share of Ten Largest

Petroleum Companies (*in percent*)

Rank	Company	1968	1969	1970	1971	1972
1	Texaco	8.5	8.3	8.1	8.4	8.1
2	Shell	8.3	8.2	7.9	7.3	7.1
3	American	7.6	7.5	7.3	7.0	6.9
4	Exxon	8.0	7.6	7.4	7.1	6.9
5	Gulf	7.5	7.6	7.1	6.7	6.5
6	Mobil	6.7	6.7	6.6	6.4	6.4
7	Arco	5.8	5.7	5.5	5.5	4.9
8	Socal	5.3	5.2	5.0	4.7	4.7
9	Phillips	4.0	4.0	4.0	3.9	4.1
10	Sun	4.4	4.3	4.1	4.1	3.9
	Total	66.1%	65.1%	63.0%	61.1%	59.5%

Source: Compiled from the *Oil & Gas Journal.*

David Attacks Goliath

The majors were in a bind. As they increased their prices to levels where they hoped to earn a decent return from marketing and refining investment, they became less competitive. Old differentials of one to two cents per gallon increased to three to five cents per gallon. The independent gasoline marketers grew rapidly since they did not have to increase their prices to make their marketing profitable. The private branders were in their element since their major brand competitors were having to reflect their much higher costs in the price at which they sold their gasoline.

As long as the independent private brand marketers and independent refiners could obtain competitive supplies of gasoline and crude oil, the majors had no way to significantly increase their own marketing and refining prices to the levels needed to improve their rate of return. Competition in the marketplace was finally doing its job. The competitive process, had it been allowed to work, would have soon sent most of the friendly neighborhood service stations the way of the "ma and pa" grocery store—into the graveyard of outdated marketing ideas.

The Onslaught of Self Service

Timing could not have been worse for the major oil companies to have decided to try to make money on the marketing and refining of gasoline. As the majors set about increasing their gasoline prices with the goal of improving their return, several independents adopted and perfected the more efficient and lower cost self-service method of marketing. The innovative self-service operators enjoyed substantial labor savings by having customers serve themselves. In addition, many self-service operators effected large operating savings from limited investment in facilities and land.

The consequence of majors trying to increase their prices while many independents were adopting the more efficient self-service method of marketing was that spreads between the major brand and independent prices increased. Customer savings of two to three cents for regular gasoline in the mid-1960's increased to five to seven cents in the early 1970's for independent self service in comparison to the major brand service price for gasoline. With this growing price spread, the majors could no longer maintain a check on the growth of independent discount gasoline marketing. The self-service gasoline revolution led by the independents, combined with traditional discounting where self-service was prohibited by law, broke the back of the inefficient and outmoded major brand method of marketing.

Readers interested in learning more about the history, growth, impact, and evolution of self service are encouraged to refer to Appendix A, "The Terrifying Growth of Self Service."

Summary

The crude oil division of the integrated petroleum companies was established as the major profit center of the industry following World War II. This was accomplished by the administration of artificially high and noncompetitive crude oil prices which drew industry profits away from downstream activities of refining, distribution, and marketing, to tax-sheltered crude oil operations. The monopoly pricing of crude oil worked to the benefit of the large integrated petroleum companies and to the detriment of independent refineries, distributors, and marketers.

From 1950 to 1970 crude oil was the "profit nerve center" of the petroleum industry. Downstream refining, distribution, and marketing operations of most integrated oil companies did not come anywhere near earning a competitive rate of return and were subsidized by huge crude oil profits. As the 1970's approached, however, the high profit was being squeezed out of crude oil both domestically and internationally. With changing conditions, downstream profitability was needed to improve the earnings of the petroleum companies.

As profits of the integrated petroleum companies sagged in the late 1960's and early 1970's, the integrated petroleum companies tried to increase their downstream prices and return on investment. However, as they did this their grossly overbuilt and extremely costly major brand system of marketing faltered and had to be supported with billions of dollars of price allowances which cut deeply into much needed profits.

The major oil companies' timing for increasing gasoline prices could not have been worse. As the majors were going up in pricing, the independent, private brand segment of the gasoline industry was perfecting and adopting a lower cost and lower price method of marketing gasoline—self service. The low price, independent, self-service method of marketing gasoline was breaking the back of the costly and outmoded major brand method of marketing. The competitive process was forcing a revolution in gasoline marketing with possible consumer savings of billions of dollars per year.

5

Design of Petroleum
Product Shortages

Introduction

The goal of the major oil companies to improve downstream profitability of marketing and refining investment was discussed in the previous chapter. Working toward this end the major oil companies tried a number of strategies. These included increasing the normal retail price of gasoline and the associated gasoline margins, adopting the self-service method of marketing that had been pioneered by the independents, withdrawing from less profitable markets and consolidating efforts in areas of strength, eliminating price-war subsidies to dealers and replacing them with a rack pricing system, and developing major secondary brand outlets.

The efforts of the large integrated oil companies to improve profitability of their marketing and refining activities were essentially useless. One of the major deterrents to the profit improvement of the large oil companies was the independent, private brand, discount gasoline marketing operations that sold gasoline to the public for less. While the major oil companies increased their prices and margins hoping to increase their profitability, the independent discount gasoline marketers enjoyed rapid growth at the expense of the branded service stations and controlled outlets of the major oil companies.

The failure of the oil companies' efforts to improve their downstream profitability in the early 1970's was related to the extreme inefficiency of their method of marketing gasoline. With a relatively low or no return on investment requirement, market representation and coverage were frequently the goals for major-brand marketing operations. This led to massive overbuilding of retail outlets. A chief executive officer of Exxon pointed out that, during the 1960's and into the 1970's, many times the number of stations needed to meet demand were built and this became a problem for the petroleum industry.

Contributing to the problem, too, has been the major oil companies' adherence to fundamentally a nonprice method of marketing. With price deemphasized, convenience through number of stations became an important competitive tool. The combination of the low return on investment requirement and the stress on nonprice methods of marketing led to the costly, top heavy and backward method of major brand marketing. As the large oil companies began to feel pressure to improve downstream profitability in the latter 1960's and early 1970's, the rate of station building began to fall.[1] Even without a shortage of petroleum products, pressures being exerted by the efficient, independent private brand method of marketing would likely have forced the major oil companies to pare to one-half or less the number of their stations. Put to the test of the marketplace, large numbers of unneeded, unjustifiable, and uneconomic major brand service stations would have been forced to close, and the "invisible hand" of the market mechanism would have done its job.

The public consequence of what was beginning to happen to major brand marketing would have been a lowering of the cost of marketing gasoline and the price at which gasoline is sold to the public. Considering that approximately 100 billion gallons of gasoline are sold per year, every one-cent reduction at the pump results in the public saving close to one billion dollars. Had the marketplace not been short-circuited by the petroleum product shortages, perhaps two to three cents could have been peeled off the 12 to 14 cents which it costs majors to make a fair rate of return on their marketing investment. Considering that the majors market around 70 percent of the gasoline sold in the United States, the annual savings to the public would have been one-and-a-half to two billion dollars per year—a sizeable savings.

Unfortunately, pressures compelling the major oil companies to reduce costs and prices or to compete have nearly evaporated with the shortage of gasoline supplies. As a result the majors have rescued their overbuilt, costly, and archaic methods of marketing gasoline to the public. Because of the structure of the industry and the limitation of supplies, the large petroleum companies have been able to divert the supplies of gasoline from the most efficient to the least efficient method of marketing gasoline—their branded outlets.

In summary, the petroleum product shortages have been used by the large petroleum companies to neutralize the competitive force for change to more efficient and logical methods of marketing gasoline to the public. This chapter will show the behavior of the petroleum industry that created the petroleum product shortage which subsequently had such a severe and noncompetitive effect. When considering the development of the petroleum product shortages during 1972, it is well to keep in mind that the U.S. was the only place in the free world experiencing shortages. In fact, throughout much of the remainder of the world, the oil companies were endeavoring to increase their sales of petroleum products.[2]

Structuring the Petroleum Product Shortages

In the late summer of 1972, initial signs of shortage appeared as gasoline supplies became very tight in the mid-continent area of the United States. For the first peace time since World War II, some private brand discount gasoline marketers were unable to purchase needed supplies of gasoline at reasonably competitive prices.

The early signs of the gasoline shortage were followed during the winter of 1972–73 by a shortage of distillate fuel oil, of which home heating oil constitutes a major component. The winter heating oil shortage led to a gasoline shortage during the summer of 1973.

As a consequence of the petroleum product shortages, schools were forced to close and airplanes had to make extra fueling stops in areas where jet fuel was more readily available. Fuel oil customers were allotted products and forced to lower their thermostats to conserve fuel. Finally, during the summer of 1973, gasoline was rationed to service stations and consumers were urged to conserve. The concern over being caught far from home with no place to buy gasoline, as happened in Denver, resulted in the driving public taking fewer and shorter trips.

The shortages of petroleum products experienced in the United States did not just happen overnight. The "shorting of production" took place several months before the publicly recognized shortage of petroleum products. The critical period that preceded the recognized petroleum products shortages was the first half of 1972, when low refinery runs created a tightness in supply of petroleum products. The refineries' operations during the final quarter of 1972 were also important, and at that time a shortage could still have been avoided. However, since production was not at a maximum, shortages of heating oils grew severe during the winter of 1972–73. Finally, the relatively low production of gasoline during the first five months of 1973 made a gasoline shortage inevitable during the peak driving season of the summer of 1973.

Underutilization of Refining Capacity

Refineries must operate at sufficient capacity to meet market demand for the different petroleum products. Otherwise, shortages of petroleum products will occur, accompanied by a fall in inventory levels. It will be shown in this section that refineries were not operated at high enough capacity to keep up with demand. This behavior by the U.S. refining industry was largely responsible for the shortage and the reduction in inventory of the basic petroleum products.

To contrast the industry's capability and its actual output, it is first necessary to define a "maximum sustainable rate of refining capacity utilization." Industry and government indicate that 92 to 93 percent of capacity utilization based upon "crude oil runs to stills" is the maximum sustainable rate of output. In nontechnical terms, "crude oil runs to stills" simply means the amount of crude oil that is processed by the distilling unit of a refinery. Through a heating process crude oil is cracked into products of different densities, which are further processed to make gasoline, heating oils, asphalt, etc.

Staff members of the Office of Oil and Gas, the statistical forecasting section of the Department of the Interior, indicated that the 92 percent rate of maximum capacity utilization takes into account normal down time. The only down time not considered is that resulting from extraordinary events which are not predictable on a regular basis for the typical refinery.

Supporting the conclusion that refineries can sustain operation at 92 to 93 percent of capacity are industry statistics showing U.S. refineries (east of the Rocky Mountains) operated during June, July, and August of 1973 at 95.4, 93.5, and 94.6 percent of capacity. The average capacity utilization of 94.5 percent during that three-month period is considerably above the 92 to 93 percent level.

Refineries located east of the Rocky Mountains (PAD District I–IV) operated at only 87.5 percent of capacity during 1972 (see Table 5–1). From January through April 1972, refinery capacity utilization was lower than for any comparable period of 1970, 1971, and 1973, as can be readily observed from Figure 5–1 and Table 5–1. For the first four-month period of 1972 refineries operated at only 84.2 percent of capacity in contrast to 90.7 percent in 1970, 87.2 percent in 1971, and 90.6 percent in 1973. The differences amount to utilizing 6.5 percentage points less capacity in 1972 than in 1970, 3.0 percentage points less than 1971, and 6.4 percentage points less than 1973.

Similarly, during the first six months of 1972 refineries east of the Rockies operated at only 85.4 percent of capacity in comparison to 90.7 percent during 1970, 87.2 percent during 1971, and 91.6 percent during 1973. Thus, during the first half of 1972 refineries utilized 5.2 percentage points, 1.8 percentage points, and 6.2 percentage points

TABLE 5–1

Refiner's Capacity Utilization East of Rocky Mountains—
Crude Oil Runs to Stills

1972 Week Ended		API Weekly				Monthly Average			
		1970	1971	1972	1973	1970	1971	1972	1973
January	7	95.0	90.2	85.6	91.4				
	14	91.0	85.3	83.3	91.2				
	21	91.0	87.5	82.7	91.6				
	28	89.8	87.3	85.3	90.8	91.7	87.6	84.2	91.3
February	4	90.3	87.0	83.5	91.7				
	11	90.9	86.2	82.7	91.3				
	18	91.3	88.2	82.2	91.1				
	25	89.9	88.3	84.8	91.6	90.6	87.4	83.3	91.4
March	3	89.1	88.7	85.7	91.4				
	10	91.9	87.3	86.7	89.7				
	17	93.8	86.8	84.2	90.6				
	24	92.3	87.9	84.9	90.7				
	31	93.3	85.4	84.3	88.6	92.1	87.2	85.2	89.9

TABLE 5-1—*Continued*

1972 Week Ended		API Weekly				Monthly Average			
		1970	1971	1972	1973	1970	1971	1972	1973
April	7	89.9	87.1	85.0	90.2				
	14	87.2	87.9	83.8	89.2				
	21	87.4	86.2	83.9	88.0				
	28	87.7	84.9	83.1	91.6	88.1	86.5	84.0	89.8
May	5	87.6	80.9	83.8	86.9				
	12	86.7	80.8	87.0	87.8				
	19	85.7	82.0	87.3	90.0				
	26	86.8	84.3	86.4	93.2	86.7	82.0	86.3	90.2
June	2	88.8	86.4	87.1	93.0				
	9	89.5	87.8	86.7	92.9				
	16	87.0	89.0	90.0	95.9				
	23	90.2	91.6	90.0	96.8				
	30	89.1	90.6	90.8	95.9	88.9	89.1	89.4	95.4
July	7	88.7	88.5	88.7	94.4				
	14	86.1	88.7	89.8	93.7				
	21	88.3	87.6	89.7	94.6				
	28	88.7	86.5	90.3	94.7	88.0	87.8	89.8	93.5
August	4	87.2	86.9	88.6	96.5				
	11	85.5	88.3	91.4	95.4				
	18	88.9	88.1	89.7	94.4				
	25	91.7	88.2	88.2	96.0	88.3	87.9	89.7	94.6
September	1	90.3	83.3	90.8	95.3				
	8	88.1	85.8	92.7	93.8				
	15	87.6	85.9	92.0					
	22	86.3	86.8	91.0					
	29	87.2	87.3	91.8		87.9	85.8	91.9	
October	6	85.5	85.0	86.5					
	13	88.7	85.9	88.3					
	20	88.3	86.8	89.6					
	27	88.2	86.0	88.8		87.7	85.9	88.3	
November	3	88.8	83.3	88.6					
	10	88.7	83.6	85.1					
	17	91.4	84.0	86.4					
	24	92.6	84.4	88.8		90.4	83.8	88.1	
December	1	85.8	89.5	91.6					
	8	86.2	85.7	89.8					
	15	90.3	86.8	89.7					
	22	89.5	86.4	90.0					
	29	88.2	86.4	92.5		88.0	87.0	90.5	

Source: American Petroleum Institute, Weekly Statistical Bulletins.

FIGURE 5–1
Refining Capacity Utilization
PAD I–IV (API)

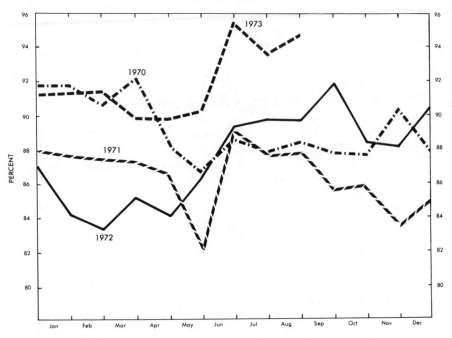

Source: Table 5–1.

less capacity than had been true for the years 1970, 1971, and 1973, respectively.

From the low level of 85.4 percent of capacity utilization during the first half of 1972, refinery output jumped in the third quarter to 90.4 percent—up almost 5 percentage points. However, during the final quarter of 1972 refining capacity utilization slipped to 88.9 percent—down by 1.5 percentage points.

The relatively low refining runs during the first half of 1972 and the decrease in capacity utilization from the third to the fourth quarter caused a drop in basic inventory levels for 1972. An industry statistic used to assess the inventory situation is days' supply of a particular product. As can be seen from Figure 5–2 and Table 5–2, the days' supply of distillate fuel oil inventories, including home heating oil for 1970 and 1971, closely track one another. However, as refining capacity utilization was reduced during the first half of 1972, the

TABLE 5-2

Days' Supply of Inventories of Distillate Motor Gasoline and Crude Oil for PAD Districts I–IV from 1970–1973

A. Distillate—Days' Supply of Inventories

Year	Jan	Feb	Mar	Apr	May	Jun	Jul	Aug	Sep	Oct	Nov	Dec
1970	41.1	36.0	34.5	39.5	52.4	66.9	86.6	100.8	101.4	96.7	85.7	62.4
1971	48.5	41.1	39.9	43.3	54.7	66.2	88.5	101.1	102.3	102.8	75.0	60.9
1972	52.3	38.2	34.0	38.3	42.0	52.1	75.9	79.1	82.6	71.5	60.7	43.2
1973	37.3	30.4	34.0	43.0	44.4	51.5						

B. Motor Gasoline—Days' Supply of Inventories

Year	Jan	Feb	Mar	Apr	May	Jun	Jul	Aug	Sep	Oct	Nov	Dec
1970	41.8	43.8	43.7	43.4	41.1	37.5	35.1	34.2	34.1	34.7	35.5	35.2
1971	40.5	43.2	44.1	41.1	40.8	35.7	33.8	33.7	35.4	35.9	35.2	35.5
1972	41.5	43.8	40.5	39.4	36.3	32.6	31.0	29.8	30.7	32.6	32.9	33.6
1973	35.9	35.3	34.2	33.2	31.1	30.4						

C. Crude Oil—Days' Supply of Inventories

Year	Jan	Feb	Mar	Apr	May	Jun	Jul	Aug	Sep	Oct	Nov	Dec
1970	24.2	24.4	24.5	25.6	26.2	26.1	25.9	24.1	23.3	24.2	24.4	24.7
1971	25.3	24.7	24.7	24.4	25.7	25.0	24.8	24.4	24.9	24.2	24.0	23.4
1972	22.7	22.0	22.3	23.1	23.2	23.4	22.5	21.8	20.8	20.7	21.2	20.5
1973	19.9	19.2	19.3	19.8	20.3							

Source: Inventory figures—Bureau of Mines Monthly Petroleum Statements. Demand figures used in calculating days' supply obtained from Bureau of Mines Monthly Petroleum Statements (advance release).

FIGURE 5–2
Days' Supply of Distillate Fuel Oil
PAD I–IV

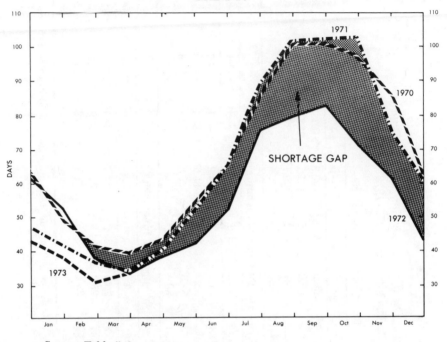

Source: Table 5–2.

days' inventory of distillate fuel oils, including home heating oil, fell sharply below the prior two years. In June 1972 distillate inventories had fallen to 52.1 days—14 days behind the levels during 1970 and 1971. By October 1972 the days' supply of distillate fuel oil had fallen more than 28 days behind the average level of the prior two years.

The days' supply of gasoline inventories is presented in Figure 5–3 and Table 5–2. Like distillate fuel oils, the days' supply of gasoline inventories for 1970 and 1971 closely track each other. As refinery runs were reduced during the first half of the year, the days' supply of gasoline inventories dropped below the level of the prior two years. However, the critical drop in inventories of gasoline did not occur until about February 1972. With the relatively low refinery runs during the first quarter of 1973, it was necessary for refineries to maximize the yields of distillate fuels to reduce the impact of the heating oil shortage. As this was done, the days' supply of gasoline

FIGURE 5–3
Days' Supply of Motor Gasoline
PAD I–IV

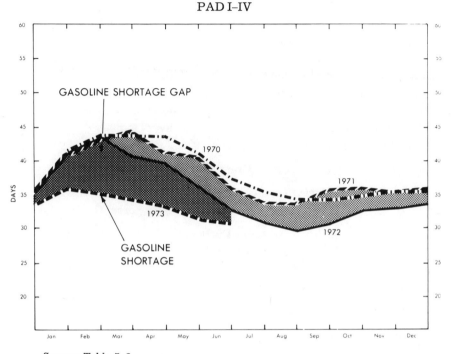

Source: Table 5–2.

inventory fell sharply. In January 1973 the days' supply of gasoline inventories fell to 35.9 days, which was five days below the average for 1970 and 1971; in February 1973 it was eight days below; and by March 1973 the days' supply level had fallen almost ten days below the average levels of 1971 and 1972. What followed was a shortage of gasoline, since inventories entering the summer months were extremely low and production was not adequate to keep up with the demand for gasoline during the peak driving season.

Another way of viewing the consequence of the low refining capacity utilization during the first six months of 1972 and the low capacity operations during the final quarter of 1972 is to contrast inventory levels with demand. Comparisons are made between 1971, the first full year before the shortage, and 1972 and 1973, the years during which the shortages developed and became widespread. The growing divergence between inventory levels and demand for distillate fuel oil in the winter heating oil shortage of 1972–73 can be observed from Table 5–3, Columns 1 and 2. By December 1972 distillate fuel oil inventories had fallen to 88.3 percent of the prior year, while demand was 118.1 percent of the base year. The inventory build-up before the heating season and production during the winter months of 1972–73 were not sufficient to keep up with demand and the U.S. experienced a severe heating oil shortage.

The gasoline supply situation was characterized as "tight" during 1972 but it did not grow into a full scale gasoline shortage until the early summer of 1973. During 1972 the average daily demand for gasoline increased 5.9 percent over 1971, while inventories trailed

TABLE 5–3

*Distillate Fuel Oil and Gasoline Inventories and Demand
for 1972 and 1973 Divided by 1971 (in percent)* [a]

	Distillate Fuel Oil		Gasoline	
	Inventory	*Demand*	*Inventory*	*Demand*
1972				
January	99.9	90.7	102.7	102.8
February	101.5	108.7	100.7	101.9
March	90.6	110.2	97.6	111.2
April	84.1	98.6	95.4	99.7
May	85.1	115.7	97.2	110.2
June	84.1	110.8	95.7	105.1
July	85.1	100.5	96.6	103.4
August	86.7	113.4	96.0	110.0
September	88.6	108.1	96.9	106.6
October	88.8	129.0	97.5	105.4
November	84.1	106.8	100.0	105.7
December	81.3	118.1	97.2	105.4
1973				
January	79.9	100.6	96.3	114.4
February	86.9	110.0	92.1	116.3
March	94.2	103.0	87.2	113.3
April	94.8	99.8	86.2	103.6
May	97.2	125.5	89.9	118.0
June	93.2	122.9	94.5	107.7
July	91.4		97.5	
August	90.5		108.6	

a. Inventory indices are based on average monthly inventories.
Source: Inventories—API Weekly Statistical Bulletin. Demand Figures—Bureau of Mines Monthly Petroleum Statement (advance release).

close behind at 97.8 percent of the previous year. The gasoline inventory picture rapidly deteriorated during the first quarter of 1973 and became widely recognized as a gasoline shortage during the summer months of 1973. By March 1973, gasoline inventories were 87.2 percent of the base year, with demand 113.3 percent of the base period. The U.S. had traded a heating oil shortage for a gasoline shortage.

Once again the reduction of gasoline inventory levels and the subsequent 1973 gasoline shortage can be related to the levels of refinery capacity utilization. During the first five months of 1973 refineries operated at 90 percent capacity, while demand for gasoline during this period increased 6.5 percent relative to 1972. The consequence of underutilization of capacity was a gasoline shortage, as demand for gasoline increased with the return of warmer weather. In June refinery capacity utilization jumped to 95.4 percent and continued for several months around 94 percent. However, this increase in capacity utilization came too late to avert the gasoline shortage which occurred during the peak summer driving season.

Refinery Operation

The major statistics on petroleum industry operations prepared by the API and the Bureau of Mines do not reveal individual company operations. However, the Texas Railroad Commission (TRC) does report some individual company information for those who elect to purchase crude oil in the state of Texas. These companies account for approximately 85 percent of refinery output east of the Rocky Mountains. Unlike the API and the Bureau of Mines, the Texas Railroad Commission reports only total runs to stills. Thus, included in the refining capacity utilization figures reported to the TRC are other inputs to the distilling unit. With the inclusion of natural gas liquid and other small inputs, refinery runs reported by the TRC are 4 to 6 percent higher than those calculated on the basis of crude oil runs to stills.

The previous analysis using the American Petroleum Institute data highlighted the relatively low utilization of refining capacity during the first four months of 1972. Now, with individual company data and aggregation of individual company data, it is possible to determine which groups of companies and which companies individually were responsible for the reduction in capacity utilization which

TABLE 5-4
Refinery Production and Capacity Utilization of Companies Reporting to Texas Railroad Commission

A. 25 Largest Refiners

Year	January 1972 less 1971		February 1972 less 1971		March 1972 less 1971		April 1972 less 1971		Weighted Average 1972 less 1971	
	% Capacity	Production B/Da	% Capacity	Production B/Da	% Capacity	Production B/Da	% Capacity	Production B/Da	% Capacity	Production B/Da
1972	90.3	8,163.7	90.5	8,199.3	90.2	8,135.9	90.5	8,131.7	90.4	8,157.7
1971	91.1	8,050.1	90.9	7,998.8	89.8	7,936.5	91.1	8,053.4	90.7	8,009.7
72 Less 71	-.8	113.6	-.4	200.5	.4	199.4	-.6	78.3	-.3	148.0

B. 10 Largest Refiners

Year	January 1972 less 1971		February 1972 less 1971		March 1972 less 1971		April 1972 less 1971		Weighted Average 1972 less 1971	
	% Capacity	Production B/Da	% Capacity	Production B/Da	% Capacity	Production B/Da	% Capacity	Production B/Da	% Capacity	Production B/Da
1972	88.8	6,217.0	87.6	6,131.9	89.4	6,240.7	90.0	6,294.7	89.0	6,221.1
1971	90.5	6,193.5	89.8	6,104.0	88.8	6,073.3	90.6	6,204.6	89.9	6,143.9
72 Less 71	-1.7	23.5	-2.2	27.9	.6	167.4	-.6	90.1	-.9	77.2

C. Next Largest Refiners

Year	January 1972 less 1971		February 1972 less 1971		March 1972 less 1971		April 1972 less 1971		Weighted Average 1972 less 1971	
	% Capacity	Production B/Da	% Capacity	Production B/Da	% Capacity	Production B/Da	% Capacity	Production B/Da	% Capacity	Production B/Da
1972	95.4	1,916.6	100.4	2,037.5	93.0	1,865.6	91.9	1,808.2	95.2	1,907.0
1971	93.8	1,856.6	94.8	1,894.8	93.1	1,863.2	92.7	1,848.8	93.6	1,865.9
72 Less 71	1.6	60.0	5.6	142.7	-.1	2.4	-.8	-40.6	1.6	41.1

a. 000 omitted.

ultimately led to a shortage of petroleum products in the United States.

For the first four months of 1972, the twenty-five largest companies reporting to the Texas Railroad Commission operated at 90.4 percent capacity, down from 90.7 percent in 1971, as can be seen on Table 5–4. Furthermore, it can be observed from Table 5–4 that the ten largest refining companies reduced their operations by .9 percentage points, to 89.0 percent in 1972 from 89.9 percent during 1971. These ten largest refineries accounted for approximately three-fourths of the production of the twenty-five largest firms reporting to the TRC. In contrast with the ten largest refinery companies, those ranked 11 to 25 increased their capacity utilization by 1.6 percentage points to 95.2 percent capacity for the first four months of 1972.

Refinery capacity utilization for the ten largest and then for companies ranked 11 to 25 for different time periods during 1972 are presented in Table 5–5. As can be observed from Table 5–5, the ten

TABLE 5–5

*1972 Capacity Utilization for Different Time
Intervals for 25 Largest Companies Reporting
to Texas Railroad Commission*

Companies 1–10	Average Monthly Capacity Utilization January–April 1972	Average Monthly Capacity Utilization 3rd Quarter 1972	Average Monthly Capacity Utilization 4th Quarter 1972
1. Humble	86.6%	96.7%	96.2%
2. American	92.5	97.8	101.5
3. Texaco	91.0	92.9	83.4
4. Shell	91.3	102.2	96.8
5. Gulf	89.6	95.2	97.6
6. Mobil	91.9	95.8	83.7
7. Arco	86.4	93.7	89.0
8. Sun	87.0	94.7	95.0
9. Sohio	88.5	91.6	97.9
10. Socal	79.5	89.2	88.5
Weighted Average 1–20	89.0%	95.5%	93.0%
Companies 11–25 Weighted Average 11–25	95.2%	96.3%	95.0%

largest refining companies operated at *only* 89.0 percent capacity during the first four months of 1972, at 95.5 percent capacity during the third quarter, and at 93.0 percent capacity during the fourth quarter of 1972. In contrast, the smaller refineries operated from 95.0 to 96.3 percent over the same three time periods. The data presented in Table 5–5 highlights the relatively low use of refining capacity of the ten largest firms in contrast to the companies ranked 11 to 25.

The refining capacity utilization of the ten largest firms for the first four months of 1972 are compared with the same period during 1971 and the data is presented in Table 5–6. Five of the ten largest refineries actually reduced their utilization of refining capacity during the first four months of the year 1972 relative to 1971. For example, Humble utilized 4.5 percentage points less capacity, Texaco 3.3, Arco 13.5, Sun 8.5 and Gulf .6 percentage points less than employed in the first four months of 1971. These five companies operated at only 88.1 percent capacity during the first four months of 1972.

As a consequence of the constrained refinery output during the first four months of 1972, the level of basic inventories including gasoline, distillates, and kerosene fell substantially below the previous year, while demand was considerably higher. For the first half of 1972, the ten largest companies held 94.7 percent of inventories of the previous year and companies 11 to 25 held 101.6 percent, while demand increased by 5 percent (see Table 5–7). For the four months of March to June 1972, after the early-year inventories had been drawn down by low refinery output, the ten largest companies held average inventories of 92.3 percent of the previous year, while companies ranked 11 to 25 maintained average inventory from March to June of 1972 of 99.6 percent of the same period a year before. It is clear that the decrease in the inventory level of the twenty-five largest firms reporting to the Texas Railroad Commission during the first half of 1972 was primary caused by the reduction in inventory levels of the ten largest petroleum companies.

The low runs of the largest refineries during the first four months of 1972 were a major contributing factor in the drawing down of fuel oil and gasoline inventories and marked the beginning of the shortage problem. With low inventories of distillate fuel oil at the beginning of the winter heating oil season, it was necessary for the giants of the industry to maintain maximum production to meet

TABLE 5-6

Change in Refinery Capacity Utilization and Production of
Refining Companies Reporting
to Texas Railroad Commission
(PAD I-IV)

A. Companies 1-10

Companies 1–10	January 1972 less 1971 % Capacity	January Barrels Per Day	February 1972 less 1971 % Capacity	February Barrels Per Day	March 1972 less 1971 % Capacity	March Barrels Per Day	April 1972 less 1971 % Capacity	April Barrels Per Day	Average 1972 less 1971 % Capacity	Average Barrels Per Day
Humble	-5.3	-40,000	-2.8	-15,000	-2.5	-12,000	07.3	-60,000	-4.5	-31,800
American	4.1	-7,900	1.7	-25,500	12.2	57,100	11.7	42,700	7.4	16,600
Texaco	-2.5	52,000	-1.4	60,600	-4.6	32,100	-4.5	31,000	-3.3	43,900
Shell	6.5	51,900	2.1	59,400	4.6	38,000	6.5	51,700	4.9	50,300
Gulf	-6.7	50,300	-3.9	70,600	2.8	120,600	5.5	142,000	-.6	95,900
Mobil	1.9	13,700	1.1	8,800	2.7	18,500	-2.1	-10,500	.9	7,600
Arco	-13.0	-70,100	-15.3	-81,300	-12.3	-66,300	-13.2	-70,400	-13.5	-72,000
Sun	-12.8	-60,700	-9.6	-46,700	-11.0	-52,200	-.6	-5,100	-8.5	-41,200
Sohio	.8	2,600	4.9	20,600	-3.2	-14,900	-2.2	-10,500	.08	-600
Socal	6.1	31,700	-5.7	-23,600	9.3	46,500	-7.0	-20,800	.7	8,500
Weighted Average Production	-1.7	23,500	-2.2	27,900	.6	167,400	.6	90,100	-.9	77,200

B. Companies 11-25

	January 1972 less 1971 % Capacity	January Barrels Per Day	February 1972 less 1971 % Capacity	February Barrels Per Day	March 1972 less 1971 % Capacity	March Barrels Per Day	April 1972 less 1971 % Capacity	April Barrels Per Day	Average 1972 less 1971 % Capacity	Average Barrels Per Day
Weighted Average Production	1.6	60,000	5.6	142,700	-.1	2,400	-.8	-40,600	1.6	41,100

TABLE 5-7

Changes in Refinery Utilization and Inventories
for Districts I–IV from 1971 to 1972

Average Refinery Utilization
1971 90.7%
1972 92.9%

Change in Refinery Utilization and Inventories by Month from 1971-1972 (percent)

	Jan	Feb	Mar	Apr	May	Jun	Jul	Aug	Sep	Oct	Nov	Dec	Average
'72 Utilization ÷ '71 Utilization	99.1	99.6	100.4	99.3	104.7	99.8	103.1	103.3	108.1	103.1	102.1	106.0	102.3
'72 Inventories ÷ '71 Inventories	102.0	101.3	92.2	94.6	94.1	93.2	94.0	94.5	95.7	92.2	92.8	90.5	94.6

'72 Inventories ÷ '71 Inventories	Jan	Feb	Mar	Apr	May	Jun		6–Month Average				Average March–June	
1. Exxon	100.3	101.8	90.2	87.9	91.2	92.0		93.9				90.3	
2. American (Pan)	86.0	87.7	78.2	86.5	93.6	83.1		85.9				85.4	
3. Texaco	111.5	111.1	98.9	103.9	102.3	106.4		105.7				102.9	
4. Shell	92.3	96.0	91.5	92.7	100.8	100.0		95.6				96.3	
5. Gulf	92.0	87.6	87.0	91.7	91.7	97.7		91.3				92.0	
6. Mobil	100.8	98.9	90.2	88.8	80.2	81.0		90.0				85.1	
7. Arco	110.5	105.5	96.5	104.9	107.4	95.7		103.4				101.1	
8. Sun	100.8	98.2	85.1	97.8	99.0	95.9		96.1				94.4	
9. Sohio	104.3	102.2	91.6	94.5	90.4	78.1		93.5				88.7	
10. Socal	104.8	99.0	82.9	85.8	91.8	87.7		92.0				87.1	
Average 1 thru 10	100.3	98.8	89.2	93.5	94.8	91.8		94.7				92.3	

11. Continental	111.4	105.4	92.2	90.1	94.3	111.0	100.7	96.9
12. Phillips	100.5	99.5	93.4	93.1	90.8	88.4	94.3	91.4
13. Union	119.9	119.2	137.6	129.5	115.0	90.6	118.6	118.2
14. Marathon	95.1	97.0	92.8	84.4	89.2	82.4	90.2	87.2
15. Cities Service	125.9	99.8	109.7	108.1	104.1	120.2	111.3	110.5
16. Getty	96.4	99.2	92.2	95.2	69.3	82.0	89.1	84.7
17. Coastal	128.6	—	72.2	123.8	77.0	123.9	105.1	99.2
18 Crown Central	117.3	139.7	115.3	125.1	94.4	106.4	116.4	110.3
19. Charter (Signal)	94.7	56.2	54.4	113.2	29.8	32.1	63.4	57.4
20. Skelly	89.5	89.3	67.1	75.9	109.9	93.2	87.5	86.5
21. American Petrofina	118.1	—	105.3	94.7	112.2	116.5	109.4	107.2
22. Cosden	128.7	117.7	102.3	180.9	124.1	104.0	126.3	127.8
23. Diamond Shamrock	84.5	76.8	79.1	92.2	101.4	100.5	89.1	93.3
24. La Gloria	118.2	—	71.2	118.9	84.6	68.9	92.4	85.9
25. Fort Worth	102.3	—	142.2	172.1	94.6	141.7	130.6	137.7
Average 11 thru 25	108.7	100.0	95.1	107.0	92.7	97.5	101.6	99.6

Source: Texas Railroad Commission.

demand. Instead, they reduced refinery runs in the fourth quarter to 93 percent of capacity, from 95.5 percent in the third quarter. With this less than all-out production, the shortage was inevitable, given the tightness of the inventory situation entering the winter.

Crude Oil Feed Stocks

To meet the demand for petroleum products refineries must have adequate levels of crude oil feed stocks to process. If refining companies are denied the necessary level of feed stocks and are forced to struggle with a tight crude oil supply, the consequences will be declining crude oil and finished product inventories. The figures and data analyzed in this section show that the refining industry was starved of adequate supplies of crude oil feed stocks, and this factor certainly contributed to the shortages of heating oil and gasoline of 1973.

Three of the most important inventories maintained by the petroleum industry are crude oil stocks, gasoline stocks, and distillate fuel oil. Crude oil is the primary feed stock processed by refineries, and gasoline and heating oil combined account for approximately three-fourths of refinery output in the United States.

These three inventories were combined on a monthly basis for the years 1969, 1970, 1971, 1972, and 1973, as shown in Figure 5–4 and Table 5–8. An analysis of these basic inventory stocks shows that they grew from 1969 to 1970 and from 1970 to 1971. It was a matter of basic stocks increasing along with increases in the level of demand. Now note from Figure 5–4 what happens to the combined inventory levels during 1972. During 1972 the basic inventory levels fall considerably below 1971, then 1970, and by May below 1969. Thus, during the first half of the year 1972, the supply situation had grown terribly tight but had not yet been recognized as serious.

The days' supply of crude oil for 1970, 1971, 1972, and 1973 are shown in Figure 5–5 and Table 5–2. As can be observed from the chart, the days' supply of crude oil feed stocks for 1970 and 1971 closely track each other throughout most of the year. However, by the end of 1971, the days' supply of crude oil had begun to drop considerably below the level of 1970. As can be seen from Figure 5–5, a large crude oil shortfall developed during 1972 and grew worse in 1973. By February 1973 the days' supply of crude oil inventories had

TABLE 5–8
Petroleum Stocks (PAD I–IV) [a]
(Crude Oil, Distillates, and Motor Gasoline)

Year	Jan[b]	Feb	Mar	Apr	May	Jun	Jul	Aug	Sep	Oct	Nov	Dec
1969	543,660	522,253	512,699	514,448	523,496	536,241	548,934	560,623	571,752	581,315	578,891	564,270
1970	545,406	532,451	542,330	526,883	540,569	547,429	553,804	566,140	583,157	600,646	606,851	601,575
1971	585,258	567,574	552,564	549,588	552,712	564,527	579,120	595,348	609,888	612,886	606,183	589,282
1972	567,302	542,857	524,871	518,569	519,541	527,256	533,511	547,233	559,456	565,476	555,963	535,320
1973	512,142	497,134	491,998	498,244								

a. Three-month average.
b. All inventories are end-of-month figures, in thousands.
Source: Bureau of Mines monthly petroleum statements.

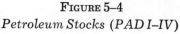

FIGURE 5–4

Petroleum Stocks (PAD I–IV)

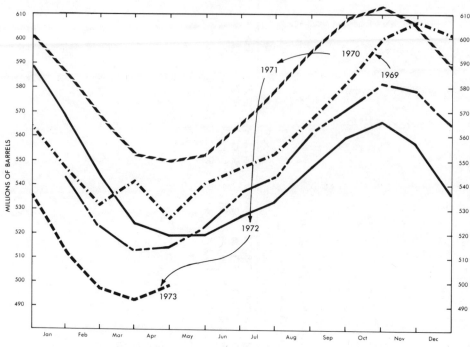

Source: Table 5–8.

fallen to 19 days from around 24.5 days for 1970 and 1971, a reduction of approximately 5.5 days' supply. Clearly, one of the significant
contributing factors to the petroleum product shortages was the
starving of certain refineries of adequate supplies of crude oil feed
stocks.

The domestic production of crude oil in barrels per day for
1970, 1971, 1972, and 1973 is presented in Figure 5–6 and Table 5–9.
Domestic production peaked in May 1972 and declined from that
point to the end of the year by about 300,000 barrels per day. With
the peaking of domestic production early in 1972, the only readily
available source of additional crude oil feed stocks was imports.

Monthly averages for daily crude oil imports for 1970, 1971, 1972,
and 1973 are shown in Figure 5–7 and Table 5–10. As can be observed
from the chart, the level of oil imports was not significantly increased
in 1972 relative to 1971. However, this was the only short-run mea-

FIGURE 5–5
Days' Supply of Crude Oil (PAD I–IV)

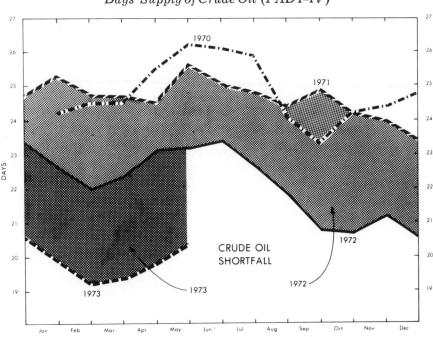

Source: Table 5–2.

sure that could be used to increase the levels of crude oil supplies for U.S. refineries to permit them to operate at high levels of capacity. While crude oil imports were increased by less than 400,000 barrels per day during 1972, it appears that oil imports were increased by close to 900,000 barrels per day in 1973 (see Table 5–10). Since the level of crude oil imports was not increased soon enough, the U.S. found itself in the midst of a very serious petroleum product shortage.

Demand, Supply, and Imports

Supply, demand, and inventories are interrelated. If demand exceeds supply, inventories will fall and give rise to a general shortage of products. The relationship between supply, demand, and inventories for fuel oil and gasoline for 1972 and 1973 in comparison with 1971 is shown in Figures 5–8 and 5–9. The shaded area between demand and supply is an approximation of the supply shortfall. As

FIGURE 5–6

Domestic Crude Oil Production (PAD I–IV)

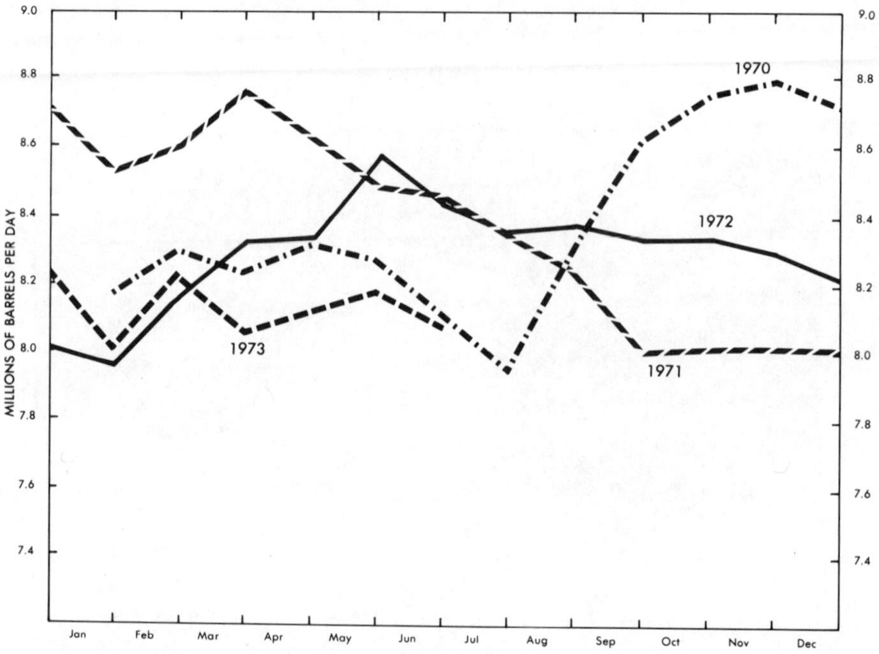

Source: Table 5–9.

can be seen, demand exceeded supply during much of 1972, inventories fell, and there was a shortage of fuel oil during the winter 1972–73.

Likewise, a similar analysis for gasoline is shown in Figure 5–9. The existence of the fuel oil shortage in the winter of 1972–73 required an emphasis on the production of fuel oil at the expense of manufacturing gasoline. Figure 5–8 shows the increased production of fuel oil in the winter 1972–73 relative to 1971 and inventory levels beginning to rise. In the same period, gasoline demand exceeded supply and inventories began to diminish, since fuel oil production was being emphasized (see Figure 5–9). Since normal off-season gasoline inventories were not built, the U.S. experienced a gasoline shortage in the summer of 1973.

With domestic production of crude oil peaking during the first half of 1972, the only way to provide an adequate level of crude oil feed stocks was to substantially increase the level of oil imports. The

TABLE 5-9
Domestic Crude Oil Production East of Rocky Mountains[a]
(in thousands of barrels)

Year	Jan	Feb	Mar	Apr	May	Jun	Jul	Aug	Sep	Oct	Nov	Dec
1970	8,189	8,299	8,236	8,323	8,252	8,102	7,944	8,316	8,616	8,771	8,794	8,714
1971	8,531	8,597	8,751	8,620	8,481	8,446	8,309	8,277	8,004	8,038	8,017	8,004
1972	7,955	8,123	8,298	8,330	8,535	8,430	8,329	8,361	8,327	8,346	8,297	8,218
1973	8,056	8,222	8,068	8,106	8,176	8,081						

a. Includes lease condensate.
Source: Bureau of Mines Monthly Petroleum Statement (advance release).

TABLE 5-10
Crude Oil Imports (PAD I–IV) 1970–1973
(in thousands of b/d)

Year	Jan	Feb	Mar	Apr	May	Jun	Jul	Aug	Sep	Oct	Nov	Dec	Average
1970	1063	1072	1090	863	931	1008	871	851	856	815	868	1042	944
1971	753	899	828	976	1010	1185	1303	1365	1309	1430	1436	1509	1167
1972	1443	1478	1379	1350	1441	1444	1516	1488	1692	1847	1612	1929	1552
1973	2105	2209	2242	2300	2403	2505	2620	2600	2600[a]	2600[a]	2600[a]		2435 (estimated)

a. Projected (in thousands of b/d).
Source: Independent Petroleum Association of America, Monthly Supply and Demand Outlook Data. Based on Bureau of Mines Monthly Petroleum Statements.

Crude Oil Imports (PAD I–IV)

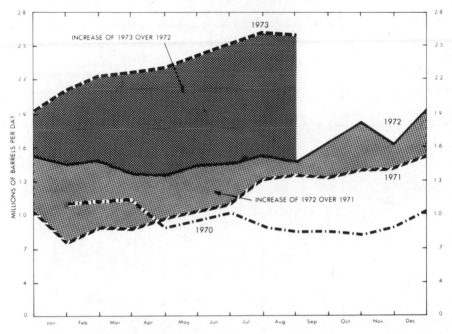

Source: Table 5–10.

responsibility and authority for managing the level of oil imports
reside with the President of the United States. The President has the
power to increase the level of oil imports, as he did twice during
1972. The major shortcoming of the administration of the Oil Import
Program was that import levels were not increased soon enough or
high enough for most refineries in the United States to have adequate
levels of feed stocks to permit high levels of operations to meet the
growing demand for petroleum products. The delays in increasing
the level of oil imports resulted in one emergency measure after
another, to offset previous errors in the administration of the quota
system. These emergency increases in the level of oil imports made
it difficult for companies to arrange for supplies and contributed to
the rapid increase of tanker rates and the costs of the importing of
crude oil into the United States.

The advice that the administrators of the Oil Import Program
received from industry was quite varied. While many of the inde-

FIGURE 5–8

*Supply, Demand, and Inventory of Fuel Oil, 1972 and 1973
in Comparison with 1971*

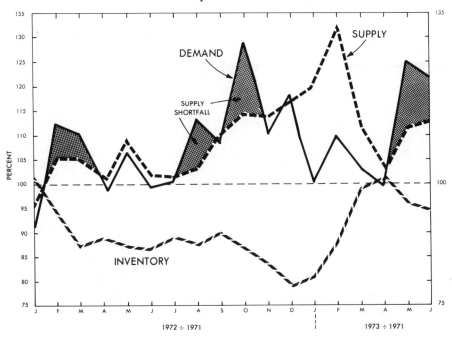

Source: See Committee Report on Fuel Shortages, p. 33.

pendent refineries and smaller integrated oil companies were encouraging higher rates of oil imports, some large trade associations
and powerful petroleum companies held a contrary view. The advice
of the latter group seemed to have been more in line with the actual
decisions of the administrators of the Oil Import Program, which for
some reason was to maintain a tight reign over the level of oil imports
as long as possible. The consequence of so doing starved the independent refineries and marketers of needed supplies of crude oil and
finished products.

The attitude of some of the larger companies on increasing the
levels of oil imports can, in part, be interpreted by their behavior
concerning the opportunity to increase oil imports in the final quarter
of 1972 by 10 percent of their prior authorized oil imports for 1972.
As Table 5–11 shows, several of the larger petroleum companies, including Exxon, Shell, Gulf, Phillips, Cities Service, and Union Oil of

California, did not use any of the additional authority to import crude
oil during the final quarter of 1972. Despite the tightness of crude oil
feed stocks in the United States, international operations with international connections ignored the opportunity to import any additional crude oil during the final quarter of 1972, although it was
needed to alleviate the shortage of crude oil in the United States.

FIGURE 5–9

*Supply, Demand, and Inventory of Gasoline, 1972 and 1973
in Comparison with 1971*

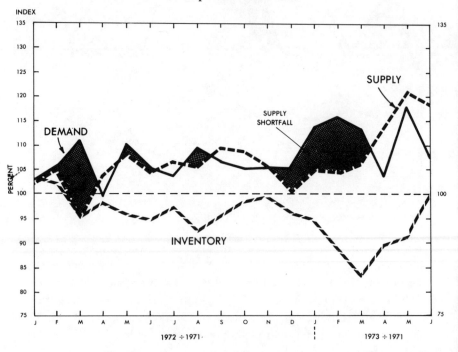

Source: See Committee Report on Fuel Shortages, p. 33.

Without these additional imports it was virtually impossible for
independent refineries and integrated petroleum companies with
low degrees of crude oil self-sufficiency to operate at high levels of
output during the final quarter of 1972 to meet the rapidly rising
demand for heating oil. Instead of importing more crude oil, some
of the giants of the petroleum industry discontinued sales of crude

oil and finished products to independent refineries and marketers, thereby forcing the shortage into the independent segment of the petroleum industry.

TABLE 5–11

Refinery 1972 Imports of Crude and Unfinished Oils
(Districts I–IV)

Amendment 46—Sec. 10 (g)

Company	Allocation 10% Advance Total Barrels	Amount Imported Prior to 12/31/72 Total Barrels	Balance as of 12/31/72 Total Barrels
Exxon Corp. (Standard N.J.)	2,051,300	0	2,051,300
Standard Oil Co. (Ind.)	1,860,405	1,119,841	740,564
Texaco, Inc.	1,783,025	1,247,247	535,778
Shell Companies	1,522,050	0	1,552,050
Gulf Oil Corp.	1,360,355	0	1,360,355
Mobil Oil Corp.	1,330,790	1,269,546	61,244
Atlantic-Richfield Co.	1,191,725	390,737	800,988
Sun Oil Co.	1,097,190	238,661	858,529
Standard Oil Co. (Ohio)	999,370	573,148	426,222
Standard Oil Co. (Calif.)	980,390	164,543	815,847
Phillips Petroleum Co.	876,365	0	876,365
Ashland Oil, Inc.	849,355	589,637	259,718
Cities Service Co.	831,835	0	831,835
Continental Oil Co.	761,025	685,011	76,014
Union Oil Co. of Calif.	712,480	0	712,480
Marathon Oil Co.	677,075	430,184	246,891
Getty Oil Co.	610,645	100,656	509,989
Amerada Hess Corp.	510,635	0	510,635
Coastal States Gas Producing	504,065	504,065	0
Union Pacific Corp.	481,435	0	481,435
American Petrofina, Inc.	479,610	0	479,610
Clark Oil & Refining Corp.	434,350	0	434,350
Tenneco Inc.	429,970	0	429,970
Crown Central Petroleum Corp.	418,655	0	418,655
Koch Industries, Inc.	401,500	0	401,500
Murphy Oil Corp.	346,750	0	346,750
Charter Company	344,560	239,156	105,404
Agway, Inc.	323,025	323,025	0
Farmland Industries, Inc.	304,775	0	304,775
Kerr-McGee Corp.	303,680	303,680	0
Subtotal	24,778,390	8,179,137	16,599,253
All Others	6,494,080	2,818,018	3,676,062
Total	31,272,470	10,997,155	20,275,315

Source: Office of Oil and Gas, U.S. Department of the Interior.

Summary and Conclusions

The preceding analysis of the operations of the petroleum industry indicates that the fuel product shortages beginning the winter of 1972–73 were largely avoidable. Sufficient refining capacity existed during 1972 and 1973. If refineries had operated nearer to their capacity levels during 1972 and during the first five months of 1973, refining capacity would have been adequate to meet demand for home heating oil and gasoline through the summer of 1973 when combined with petroleum product imports.

The heating oil shortage for the winter of 1972–73 (the final quarter of 1972 and the first quarter of 1973) was a direct result of the relatively low level of refining capacity utilization during 1972. Refinery capacity utilization during the first four months of 1972 was actually below that of the prior year. This resulted in heating oil inventories at the beginning of the final quarter of 1972 being more than 11 percent below the 1971 level. With relatively low heating oil inventories entering the winter heating oil season, it was necessary for U.S. refineries to operate at high levels of capacity during the final quarter of 1972 to meet a demand that was up 19 percent over the previous year. However, since the petroleum industry did not operate at maximum capacity during the final quarter of 1972, the heating oil shortages for the winter of 1972–73 became inevitable.

An analysis of individual company operations, using Texas Railroad Commission information, showed that several of the larger U.S. refineries were primarily responsible for the abnormally low level of refining capacity utilization during the first four months of 1972. As a result, these larger petroleum companies also drew down their heating oil and gasoline inventory levels. Furthermore, the largest petroleum companies reduced their capacity utilization in the final quarter of 1972 from the third quarter level when all-out production was needed to avoid a shortage.

Similarly, the below-capacity utilization of refineries during the first five months of 1973 gave rise to the gasoline shortage during the summer of 1973. First in the case of heating oil, and later in the case of gasoline, off-season inventories were not built up as they had been in the past and output during peak heating oil and gasoline seasons could not keep up with demand.

During 1972, particularly in the second half of the year, several

U.S. refineries were short of needed crude oil. Domestic production of crude oil peaked early in 1972, and it became necessary for oil imports to be rapidly increased in order to provide all refineries with necessary feed stocks. Regulation of the level of oil imports was the responsibility of the executive branch of government. The advice given to oil policy administrators concerning increasing the level of oil imports was quite varied. Some powerful associations and companies advised against increasing the level of oil imports, while other companies—lesser majors and independent refiners—urgently pleaded with the executive branch of government to increase the level of oil imports.

The executive's delayed decisions to increase the level of oil imports, and the nature of the decisions themselves, effectively starved the U.S. refining industry of feed stocks needed to operate at higher levels of capacity output. Furthermore, some of the larger oil companies did not utilize the authority they were given to import more crude oil in the final quarter of 1972. Thus, inadequacy of crude oil feed stocks, which was a product of both government action and the decisions of some large oil companies not to increase imports, contributed to the beginning of the petroleum product shortage.

6

Nixon Administration
Contribution to the Shortages*

The increasing dependency of refineries on crude oil imports for processing petroleum products has made government import policy very important to the petroleum industry in meeting market demand. The executive branch of the federal government has the responsibility and authority for regulating the level of crude oil imports. Crude oil imports, taken together with domestically produced oil, provide the fundamental feed stock for processing by U.S. refineries into petroleum products like gasoline, fuel oil, etc. The great bulk of crude oil feed stock used by refineries east of the Rocky Mountains (an area referred to as PAD Districts I–IV) has historically come from domestic production of crude oil. When the Mandatory Oil Import Program was established by executive order of the President of the United States in 1959, the level of crude oil imports was initially set at 12.2 percent of domestic production. Until 1970 the level of crude oil imports was maintained at 12.2 percent of domestic production.

"Balancing Wheel"

Texas and Louisiana are the two big crude oil producing states, and in recent years they have accounted for approximately two-

* Much of the information presented in this chapter was obtained from the Senate Permanent Subcommittee on Investigation inquiry into the fuel shortages during the fall of 1973.

thirds of the domestic production of crude oil east of the Rocky Mountains. Since the productive capability of these two states exceeded demand for crude oil, production was regulated for years at below full capacity—a procedure called demand prorationing.

Texas and Louisiana, with their spare productive capacity for crude oil, acted as the "balancing wheel" for supplying crude oil needed for processing by U.S. refineries. For years Texas and Louisiana established monthly levels of production at something under full capacity to insure adequate supplies of crude oil feed stocks for refineries. Commissions controlling production in Texas and Louisiana added the projected level of oil imports and the estimated level of crude oil production from some of the lesser producing crude oil states with unregulated production. Then the amount of production by Texas and Louisiana was established that would supply adequate stocks of crude oil as required by the U.S. refineries.

In the early months of 1972, Texas and Louisiana commenced to produce what they indicated was maximum capacity.[1] This was necessary since the level of oil imports had not been increased enough in 1972 to maintain some spare capacity in Texas and Louisiana. When Texas and Louisiana were forced to maximum production, the "balancing wheel" of crude oil supply was destroyed. The responsibility for maintaining adequate supplies of crude oil then switched from the state regulary agencies—in Texas and Louisiana—to the Oil Import Program and more specifically to the executive branch of the federal government, which had the responsibility for regulating the level of oil imports.

Governmental Options

With domestic production of crude oil at a maximum it became imperative for the Nixon administration to increase the level of oil imports to insure that domestic refineries would have an adequate level of crude oil supplies coming from domestic production and imports. There were three courses of action open to the executive branch of the government. One possibility was to increase the level of oil imports sufficiently to permit Texas and Louisiana to have a mild surplus of productive capacity. These states could then perform their historic function of setting monthly levels of production which would enable most refineries to obtain necessary supplies of crude oil for processing into petroleum products demanded by the marketplace.

A second possibility was a "filling-the-gap" supply versus demand approach. Oil imports would be increased so that when combined with domestic production there would be sufficient supplies of crude oil for refineries. This method had been followed since the beginning of the Oil Import Quota Program in 1959 in the western part of the country (PAD District V). The "supply versus demand gap" approach was used in the western U.S. for regulating oil imports, instead of the 12.2 percent ratio of imports to domestic production employed east of the Rocky Mountains, because domestic productive capability relative to demand was more limited in the West.

A third possibility was to remove all restrictions on oil imports so that foreign crude oil could be imported as needed to meet demand. The long-run effect of this proposition was that domestic and international crude oil would eventually compete with one another —a condition that had not existed since the beginning of the Mandatory Oil Import Program in 1959.

The first strategy, restoring a mild surplus to Texas and Louisiana, placed responsibility for maintaining adequate supplies of crude oil on the state production regulatory agencies of Texas and Louisiana. The second approach, of regulating the level of oil imports to "fill-the-gap" left by domestic production, placed the responsibility for meeting the demand for crude oil on the executive branch of the federal government. The third approach of removing all restrictions on crude oil imports left the relative quantity of oil imports to be determined by international market forces.

The federal government opted for the second approach of trying to import enough oil to fill the gap left between domestic production and demand for crude oil. The chief legal counsel of the Office of Emergency Preparedness (OEP) vividly characterized this policy as one of "walking the tight rope"—very hard to do and very risky—a wrong move in either direction would be disastrous. As shown in the previous chapter this effort failed and certain U.S. refineries were "starved" of crude oil needed to operate at levels required to meet demand. As a consequence of the inappropriate oil import policy decision made during 1972, President Nixon was finally forced to abandon the oil import quota system on May 1, 1973.

In this chapter and the one following, the relationship between government and business will be reviewed along with the government decisions which led to creating a shortage of crude oil and in turn a shortage of petroleum products.

Level of Crude Oil Imports for 1972

President Nixon authorized an additional 100,000 barrels per day of crude oil imports for 1972 east of the Rocky Mountains (PAD District I–IV). In 1970 and again in 1971 the level of oil imports had been increased by an average of 100,000 barrels per day over the 12.2 percent ratio of crude oil imports to domestic production. With this additional allocation the level of oil imports was established at 1,550,000 barrels per day (1,250,000 from the 12.2 calculation plus 300,000). Oil imports of 1,550,000 barrels per day accounted for 15 percent of the input to refinery operations.

The Oil Policy Committee, chaired by General George A. Lincoln, decided to accept the recommendation of the working committee for establishing the level of oil imports for 1972. This "pragmatic approach" was to model the 1972 import program on the 1971 plan. According to the minutes of the Oil Policy Committee meeting of November 17, 1971, *"it is recommended* that the 1972 program be modeled as closely as possible on the 1971 program in terms of quota level and allocation systems." The minutes of the Oil Policy Committee indicated that "by 1974 or 1975 it would be necessary" to allocate overseas imports on the basis of supply/demand gap, but that "we need not make it now."

While the no-change import quota for 1972 was adopted, a respected staff economist of the OEP had prepared a report indicating that domestic production might peak early in 1972. The meaning of this report, if it was correct, was that the level of oil imports should have been increased by several hundred thousand barrels at the beginning of 1972. Information collected in the course of the government investigation on the shortage revealed that top OEP officials concerned with oil matters were not even aware of this pessimistic projection until signs of crude oil shortages became very clear in early 1972. In addition, the trade press had carried some pessimistic forecasts for domestic production which supported the need for increasing the level of oil imports at the beginning of 1972 by more than 100,000 barrels per day, but these also seemed to be ignored. Further information from the government study of the shortage showed that responsible people in the OEP were extremely busy with other important matters, such as the price freeze and national emergencies, and unfortunately oil matters got short shrift.

The OEP record revealed only limited sampling of the industry's attitudes about the level of oil imports for 1972. In a memorandum dated November 5, 1972, for General Lincoln the director of the Office of Emergency Preparedness presented the following information:

> Mr. Bennett and I [A. A. Smith] both met with S. G. Stitles, Vice-President of Shell, and one of his assistants. Both men impressed me as they have in the past, as being particularly knowledgeable regarding the MOIP. . . .
>
> . . . Both Shell and Humble did not seem to be particularly concerned about the next year's import level—they both assume it will increase another 100,000 b/d. . . .

As it turned out, Shell and Humble did not face any serious problems with crude oil supplies during 1972, although in contrast several of their competitors did.

Gene Morrell, director of the Office of Oil and Gas in the Department of the Interior, wrote a letter to General Lincoln dated November 22, 1971, concerning the consequences of forecasting higher domestic production of crude oil than could be achieved. Should the forecast for domestic production be too high, Morrell warned, it would be necessary to increase the level of oil imports. Some spare capacity could be maintained in the states of Texas and Louisiana, in this way, so that they could continue to act as "the balancing wheel." Morrell wrote:

> flexibility is usually provided by changes in State allowables, notably in Texas and Louisiana. We believe that these States will continue to act as "balance wheels" again in 1972. If, however, major departures from the forecast indicate the need for increased supplies beyond the capabilities of the domestic producing industry, supplemental quotas should be authorized at a subsequent date.
> This flexibility should be retained. . . .

The Crude Oil Shortage

Given the high degree of vertical integration of several of the large petroleum companies, it was possible to forecast that the independent refineries and crude-poor integrated oil companies would be the first to feel the pressure of a shrinking supply of crude oil. This was simply because if a shortage of crude oil was permitted to develop, those owning and otherwise controlling the crude oil would claim it and the heavy net purchasers of crude oil would be left out in the cold. As an analysis of the forthcoming documents will show, this is exactly what happened. Furthermore, it became increasingly clear that unless the executive branch of the federal government took appropriate steps to increase the much needed supply, the government's inaction would be destructive to the independents and lesser integrated petroleum companies.

Louisiana Production

At the meeting of the Louisiana Department of Conservation on January 7, 1972, several oil companies complained that the crude oil supply situation was growing critical; hence, they were requesting increased production from the state. Texaco stated, for example, "During the month of January we were unable to purchase condensate crude anywhere to cover our shortage. We were able to buy some but we were short (by) some half a million barrels."

Chevron stated, "We're in the market, we've been in the market since about the 8th or 9th of December because of some changes in our schedules, we have been unable to purchase the crude we want and need and it would take a sizeable increase in the nominations, we feel, to produce enough crude to give us what we need to buy. . . ."

Clark Oil, one of the larger independent refineries with little of its own crude oil production, presented its difficulties in securing needed crude oil supplies. A representative of Clark Oil stated, ". . . I have surveyed the market and . . . been trying to buy crude for some time. . . . I don't . . . get much encouragement even if you give a significant increase for February."

Members of the Louisiana Department of Conservation pointed out the decreasing capability of that state to produce more crude oil.

Commissioner Menefee said: "The experience of late reveals that we are losing about 15,000 barrels per day per month as a result of increased deficiencies of wells with time."

Commissioner Menefee went on to state that it was important to maintain some spare capacity within the state of Louisiana in case of a national emergency. For this reason it was unlikely that Louisiana would permit crude oil production within the state to go to maximum capability. Menefee argued,

> Louisiana and the nation at all times need the capacity to produce oil above current needs, as insurance in the event of a national emergency. The question for us today is what constitutes a reasonable and necessary reserve capacity? Since Louisiana represents a sizeable portion of the nation's reserve producing capacity, it becomes more important from a national defense standpoint that we should not completely eliminate our ability to produce more, but rather preserve the substantial excess capacity and hope that additional drilling would commence to increase the reserves.

John Ricca, assistant director of the Federal Office of Oil and Gas, wrote to Gene P. Morrell, director, on January 20, 1972, concerning Louisiana's decision to withhold some productive capability on national security grounds. He stated:

> Louisiana's unilateral action to put a ceiling on production allowables tied to estimated productive capacity and justified on National Security grounds brings into the scene several new dimensions:
>
> (1) The State is unilaterally assuming legal or constitutional authority to determine an emergency resource stockpile.
>
> (2) It is setting a precedent on ratio or quantity to be shut in (10 percent to 15 percent).
>
> (3) This is outside the scope of "conservation" or "market demand" or "import proclamation."
>
> (4) If this goes unchallenged, it's assumed California, Texas, New Mexico, Oklahoma, et al. (or all Compact members) could do the same.
>
> (5) If they did, oil imports and quotas would have to be increased appreciably and immediately.
>
> (6) It raises a question of additional "restraints" on free enterprise on state lands as opposed to Federal lands where freedom prevails except for "conservation, safety," and related reasons.
>
> A complete probe of this should be made to determine whether the national interest is really best served by such action in one state only.

Production from Texas

The looming shortage of crude oil was underscored by the annual president's meeting of the Texas Railroad Commission on March 14 through 16, 1972. At this meeting the Commission went to a 100 percent market demand factor for the month of April 1972, or, in other words, to produce at what was considered to be the maximum efficient level. This was a particularly significant development in that it represented the first time since World War II that Texas found it necessary to produce at 100 percent of its capability to meet demand. No longer could Texas be relied upon to make monthly adjustments in the rate of production to meet the needs of U.S. refineries for processing purposes.

At this meeting Harry Bridges, president of Shell Oil Company, emphasized the diminishing capability of the U.S. to meet demand for crude oil and recommended an immediate increase in the level of oil imports. He said, "The federal government should immediately issue an additional *annual* import quota of 200,000 barrels per day with full realization that a further increase may be necessary later in 1972."[2]

The impact that a crude oil shortage might have on independent refineries and certain major oil companies that produce little of their own crude oil was discussed by Bonner Templeton, General Manager, North American Division, Crude Oil Department, Mobil Oil Company. He talked of the tightness of crude oil and was quoted as having said, "Without an additional quota, a number of refiners will find it difficult to get an adequate supply—particularly those who buy crude on the spot market."*

To see that this did not happen, Templeton suggested, "It may be desirable to increase imports in order to increase stock and to enable Texas and Louisiana to return to a position of modest spare capacity."[3]

The actual minutes of the mid-March 1972 meeting of the Texas Railroad Commission contained the following comments from three

---

* Spot market refers to purchasers of crude oil who do not have contractual agreements with suppliers and who purchase at the best price from the sources available at a particular point in time. However, it was not just the so called spot purchasers that were being cut off, but also, those that have purchased on an "evergreen" basis—month to month, but year after year.

oil company executives who warned of a "fuel crisis" in the latter
part of 1972 unless crude oil imports were significantly increased.

> Mr. Oscar S. Wyatt, Jr., of Coastal States . . . attacked unnamed major
> companies for failing to let independent refineries have the crude they
> need. He said the major companies have no lack of crude for their runs.

> John K. McKinely, president of Texaco, said they have made an
> analysis of production from all states in PAD Districts I–IV and feel
> that production will probably decline in 1972. As a result crude short-
> ages are firmly indicated by the fact that crude stocks are 18 million
> barrels below desired levels in the region. . . .

> W. S. McConnor, Vice President of Union Oil of California, said that
> there is an impending shortage of crude in the last half of the year be-
> cause at current rates all import licenses will be used up by Septem-
> ber. He said that the industry needs to make the government and the
> people aware of this situation *so that there will not be a fuel crisis in
> the last quarter of 1972.* (Emphasis added.)

The suggestion made by the Mobil representative to increase the
level of imports to permit mild spare capacity of crude oil to return
to Texas and Louisiana was the same general recommendation that
had been made in the fall of 1971 by Morrell. Templeton of Mobil
and Morrell were both suggesting a means by which the government
could take action to see that crude oil shortages would not occur that
would be particularly harmful to the independent sector of the refining industry.

Looming Crude Oil Shortage

The growing shortage of crude oil in the United States was emphasized by a staff report entitled "U.S. Heading for a Close Shave
with Crude Oil This Year," *Oil & Gas Journal,* February 14, 1972.
The article stated that there existed barely enough crude oil for all
in 1972 without a supplemental boost in oil imports and that at that
time almost no spot crude oil was available for purchase. In addition,
crude oil was being imported at a high rate during January 1972
because of the tightness of supply of domestic crude oil and that the
import quota was being used up at a faster rate than it had been in
previous years.

The *Oil & Gas* article also mentioned that crude oil stocks were
being drawn down as a consequence of the tightness of crude oil

supply. Crude oil supplies were at 21.8 days of demand in January 1972, believed to be the lowest level since World War II. The article told of the vanishing import tickets and suggested that the consequence might be refiners having to reduce runs and to draw down crude oil stock. Illustrating the tightness in terms of crude oil availability, one refinery was quoted as saying, "If I had to buy 10,000 b/d in spot crude I wouldn't know where to start looking. . . ." Quoting another oil executive, the *Journal* said, "If everyone is telling the truth, I don't think there are two companies who have all the crude they would like to have between now and July 1." According to general reports most experts interviewed felt that President Nixon should authorize an increase in the level of oil imports by 200,000 barrels per day. More imports might be needed should demand increase faster than projected.

Pleas for Increased Imports

Beyond the public statements made in March and April 1972, several petroleum companies directly communicated their concern about the inadequacy of crude oil supply to the Office of Emergency Preparedness, whose chairman had primary responsibility for advising President Nixon on oil import policy. Those companies pressuring the government for increased imports were largely the independent refineries and crude-deficient integrated oil companies. While some major oil companies had warned of the prospects for crude oil shortage, they did not make direct appeals to responsible officials for more imports. Perhaps the giants of the industry recognized that if there was enough, but not too much, shortage it would work to their advantage. The marketplace could be cleaned up—products kept from the rapidly growing price marketers—and an economic squeeze could be exerted on independent refineries and integrated companies with a low degree of crude oil sufficiency. What follows are some of the statements submitted to the OEP that urged increased oil imports.

Clark Oil Company, an independent refinery and large net purchaser of crude oil, wrote the Office of Emergency Preparedness on February 21, 1972. Clark Oil recommended an increase of 350,000 barrels per day in the level of oil imports for Districts I–IV for 1972 to prevent a severe crude oil supply crunch. In addition, Clark emphasized in its statement that, if there were a shortage of crude oil,

it would fall most heavily on the crude oil-deficient independent refining segment and could literally destroy this segment if proper
government action was not taken. The final paragraph from this
letter summarized:

We think that the National Petroleum Council forecast of the need of
some 500-600,000 B/D of additional offshore imports for Districts I–V
during 1972 is very realistic. Due to the relatively small increase of
100,000 B/D for Districts I through IV in 1972, which is certainly inadequate, coupled with the Bureau of Mines forecast of a 6.6 percent increase in demand for petroleum products in 1972, it will be necessary to
increase these quotas by another 350,000 barrels for the entire year to
prevent an industry wide crude oil supply deficit of such magnitude as
to curtail refinery runs by the industry. *This curtailment would fall*
most heavily on the crude deficit independent refining segment and
would literally destroy these independent refineries if no action is
taken.

Standard Oil of Ohio sent a letter to General Lincoln dated
March 13, 1972, concerning the crude oil shortage that was developing. Sohio pointed out that there was a general shortage of crude oil
and that it was particularly severe on the big net purchasers of
crude oil, of which Sohio was one. The shortage of crude oil had
already forced Sohio to curtail its refinery runs. As a consequence
Sohio recommended an immediate increase in the level of oil imports
east of the Rocky Mountains by at least 300,000 barrels per day for
the full year of 1972. Excerpts from the Sohio letter are presented
below.

Major companies normally counted on as sellers of domestic crude are
themselves aggressively seeking crude to buy. Meantime, many refiners
who have traditionally been net crude purchasers find themselves
substantially short of supply commitments to cover their 1972 requirements. Sohio is one.
      . . . With no prospects for additional domestic crude purchases we
have already curtailed crude runs sharply at two of our refineries.
Additional cutbacks, and probably shutdowns, will be necessary soon if
no additional supplies are forthcoming, as the deficit worsens in coming
months.
      In view of our outlook, we are compelled to recommend that steps
be taken quickly to increase the level of permissible crude oil imports
into Districts I–IV by at least 300,000 B/D for the full year 1972 above
the approximately 1,200,000 B/D presently licensed

The American Petroleum Refiners Association met with officials of the Office of Emergency Preparedness on March 8, 1972, and sent a letter with additional information dated March 13, 1972. This association represents the smaller independent refineries located primarily east of the Rocky Mountains. The association representatives felt that a serious shortage of crude oil existed in the United States, that the shortage was growing worse, and that the effects of the shortage could be particularly harmful to the independent refinery. This harm would occur since many of the independents were buying crude oil from large petroleum companies that would first look after their own needs if a shortage of crude oil were to become severe in the United States. During the meeting, association representatives said:

> Published statistics present conclusive evidence that additional import licenses should be issued now to prevent a supply catastrophe over- taking small business refiners. . . . Issuance of licenses now would pro- duce normal and orderly programming for the balance of this year.

The association urged that the level of oil imports in Districts I–IV be increased by 500,000 barrels per day for 1972.

Ashland Oil Company, the largest of the independent refiners in the United States, met with the Office of Emergency Preparedness on April 4, 1972, to request an increase in the level of crude oil imports by 300,000 barrels per day. A staff economist for the OEP corrected the forecast made by Ashland according to the projections of the OEP and concluded that there was going to be a shortage of "only 24,000 barrels per day rather than 300,000 barrels per day." Ashland's projection was very close to being correct.

Coastal States is the second largest independent refining company. Oscar S. Wyatt, Jr., chairman of the board of Coastal, wrote General Lincoln on May 3, 1972, about the severe problem that Coastal Oil was having in obtaining crude oil needed to run its large Texas refinery.

> Coastal's refinery in Corpus Christi, Texas, has a capacity of 135,000 barrels per day. During January, February, and March in 1972 Coastal has utilized all of the domestic and foreign crude oil feed stock avail- able to it, and only by utilizing all of our 1972 import allocations during those three months and by utilizing all of the domestic production

available, have we been able to obtain only enough crude oil feed stock to operate the refinery at approximately 100,000 barrels per day. We have no import allocation for the rest of the year. There is not enough domestic production available to us to continue at the 100,000 barrel level.

Wyatt further explained that when the Mandatory Oil Import Program was ordered on February 27, 1959, conditions were quite different from those existing by 1972. The Mandatory Oil Import Program "was not intended to reduce imports to a point where domestic refining capacity could not be utilized." Wyatt concluded his letter by urging that oil import policies be changed to permit substantial increases in the level of oil imports to meet demand and the needs of U.S. refineries.

I strongly urge you to consider the facts as they exist today rather than as they existed in February and March, 1959, and to take the action necessary to allow additional imports, particularly to those refineries who are deficient in domestic crude and who are now excluded by existing regulations from importing sufficient crude to meet their demands.

Others Discount Need for Additional Imports

Humble Oil Company, a subsidiary of Standard Oil of New Jersey (Exxon), the world's largest oil company, informed the Office of Emergency Preparedness in March 1972 that "there was no need for additional imports in 1972" as far as Humble was concerned. "Humble at present is of the opinion that the short-fall in Texas production is a reflection of market conditions and not truly a reflection of the decline in production capacity, if any."[4] The Humble representative went on to say that the company had crude oil to sell but had had "no takers in recent weeks." The Humble representative left the impression with the assistant director of OEP that "the crude market seems to be easing."

General Lincoln, in a memorandum to Peter Flanigan, Nixon's top energy advisor, stated that opposition would be voiced by the IPAA (Independent Petroleum Association of America) and other groups which believed "there is enough potential production to meet our crude oil demand this year."[5] The assistant director of the OEP, Elmer Bennett, wrote in the margin of the memorandum:

Any reserve capacity is small at best. Consequently, I think we are faced with increasing imports even if the majors (or others) are withholding some oil from the market. We should be able to defend our action either way; if there is some capacity available or if there is not.

The assistant director for energy analysis prepared a memorandum for General Lincoln and indicated "the subject of my memorandum could well have been 'the emperor's clothes.' "[6] What he was referring to was that the government continued to try to establish the level of oil imports based upon an unreal assumption—that America could produce, under existing circumstances, all the oil it needs.

First Emergency Increase in Oil Imports

The level of oil imports was increased by 230,000 barrels per day by a Presidential Proclamation on May 11, 1972. The increase fell between the high and low estimates presented by industry and the OEP staff. In a lead article in the *Oil Daily,* entitled "Washington Changes Signals on Crude Import Plan" dated May 4, 1972, the question was raised as to why the level of oil imports was not going to be increased by more than 230,000 barrels per day. The writer of the article noted: "Some officials have argued that it would be better to err on the side of an [excessive] increase in the import level—if it later turned out to be excessive—in order to have spare capacity in Texas and Louisiana." It was further noted in the *Oil Daily* article that the "decision makers"—Peter Flanigan and OEP Director George Lincoln, Nixon's Oil Policy Committee chairman—apparently did not agree with this thesis. In trying to understand the reasoning of these men, the author hypothesized that, by having a tight situation, the government might be able to "dramatize" the energy problems of the United States. The article stated: "Their reasoning is hard to fathom. Some ascribe it to 'caution' in not permitting 'excessive' imports, or a possible move to 'dramatize' the nation's energy pinch."

The role of Peter Flanigan in the decision-making process was further described:

Flanigan is aware of what is involved in holding down the import increase and is counting on all-out production to meet oil demands during the balance of 1972. . . .

It is believed both Menefee and Texas Chairman Brian Tunnell are concerned about production of their state's oil fields at formula capacity production. . . .

As previously discussed, Texas and Louisiana had acted as the "balancing wheel" of crude oil supply as needed by U.S. refineries. To restore some mild surplus to Texas and Louisiana so that they could resume their prorationing activities would have required an increase in the oil import levels of approximately 400,000 barrels per day, instead of the approved rate of 230,000 barrels per day, for 1972. According to minutes of the Oil Policy Committee meeting, Flanigan had opposed a major increase in the level of oil imports. "Mr. Flanigan questioned that we wanted to raise the level of oil imports to such an extent that the States would move back into market demand prorationing."[7] The return of some surge capacity to Texas and Louisiana had been recommended by the director of the Office of Oil and Gas and by a Mobil Oil Company representative. In addition, one of OEP's economists had recommended the same course of action.

Continued Crude Oil Shortages and Forecast of Product Shortages

The Presidential Proclamation of May 11, 1972, announced that the level of oil imports was to be increased by only 230,000 barrels per day, while some industry representatives had requested substantially greater increases in the level of oil imports. In fact, an in-house OEP report showed that it might be necessary to increase the level of oil imports by 360,000 barrels per day and one segment of the petroleum industry had called for increases in the level of oil imports by 500,000 barrels per day.

In this section it will be shown that once again the oil policy administration in the executive branch of the federal government failed to increase adequately the level of oil imports and another emergency increase became necessary. Furthermore, it will be seen that the crude oil shortage began to affect availability of finished products.

Many of the same companies that had called attention to the crude oil shortage early in the year once again felt it necessary to plead for additional crude oil imports. Representatives of the Clark Oil Company met with the staff of the OEP at the end of July 1972 to explain that, even with the additional 230,000 barrels per day allocated by the OEP, they foresaw a "supply short-fall of 272,000 barrels per day during 1972."⁸ Clark predicted that, without additional crude oil imports, the company would have sufficient crude oil only through October 1972. This would be followed by a small shortage in November and a large shortage in December, requiring reduced refining output.

As with some of the earlier predictions of shortage by the independent refiners, one of the staff economists for the OEP, with previous major oil company background, questioned the forecast. After adjusting the forecast he concluded that there would actually be a surplus rather than a deficit of crude oil available for the remainder of 1972. Once again, the facts proved the economist to be incorrect.

Coastal States met with the staff of the OEP on August 11, 1972. Coastal informed the OEP that they had used almost all of their oil import quota for 1972. As a consequence, "they have notified all customers by letter that across-the-board prorationing at about two-thirds of demand will be in effect for the remainder of the year."⁹ Coastal went on to explain that they had tried to purchase crude oil from several sources, but this effort had been to no avail. A couple of weeks later Oscar Wyatt, chairman of the board of Coastal States, met with General Lincoln to discuss the critical shortage of crude oil that Coastal States was facing. Wyatt informed General Lincoln and staff on August 31, 1972, that the shortage was most acute with crude-deficient refiners, of which Coastal was one of the largest. As a result of the difficulty in obtaining supplies,

> Coastal has lost over 3,000,000 barrels a run in the first 8 months of 1972. Refinery runs, at approximately 73 percent in August, were run at 64 percent in September according to latest supply figures.

Wyatt recommended that imports be increased to a level which would restore some surplus productive capacity to the states of Texas and Louisiana so that independent refineries and other crude-deficit refineries could obtain supplies of crude oil. He said that the import level should be increased by "an additional 400,000 or 500,000 barrels per day. This will permit allowable systems in Texas and Louisiana to reassert themselves and provide a surge tank for unanticipated increase in demand."

This is the same policy recommendation regarding the Oil Import Program which had been made by several other people in recent months.

The continued shortage of crude oil being experienced by Sohio was presented in a letter to General Lincoln dated August 17, 1972. The letter told of the "traditional, evergreen" crude oil supply contracts with major oil companies being cancelled. These were not spot purchases of crude oil—a term often used in a disparaging way. "Spot purchases" usually infers past bad business practices resulting in some refineries having trouble getting crude oil later. Furthermore, it was explained that the tight crude oil supply was working its way down to product sales and historical suppliers of products were no longer making them available to their regular customers.

Sohio characterized the inadequate supplies of crude oil as a "bottom of the barrel sickness"—a malady affecting the U.S. refining industry. Sohio's recommendation was that there should be at least 200,000 additional barrels per day of crude oil imports authorized on an annual basis and that this decision should be made immediately so supplies could be negotiated and arrangements made for long-term and more economical transportation.

On September 1, 1972, Standard Oil of Indiana wired the Office of Emergency Preparedness that there would be an overall shortage of approximately 600,000 barrels per day of crude oil for September through December 1972. Standard indicated that it was "experiencing unprecedented difficulty in acquiring sufficient crude or products to supply present customers' needs. . . ." To meet strong market needs Standard believed that U.S. refineries would need additional imports of crude oil.

In a letter to the OEP dated August 30, 1972, a representative of the Sun Oil Company presented several dismal facts relating to the shortages of crude oil and finished products that currently existed in the United States. In the letter, it was pointed out that "we do not have any uncommitted domestic crude left. I am sure you know that we cannot buy any." It was recommended that the level of crude oil imports be increased from 100,000 to 150,000 b/d for the whole year of 1972.

In latter July and early August representatives from Exxon, Mobil, and Citgo met with representatives of the OEP. No mention was made in the memos pertaining to these meetings of a need to increase the level of crude oil imports. In contrast to the statements of crude-deficient refineries, the Citgo representative was quoted as saying

> . . . oil import quotas were adequate for the remaining portion of 1972, although there might be some spot shortages affecting some individual refineries. He said that shortages would primarily affect those refineries which purchased primarily on a spot market and have avoided long term commitment.[10]

Forecast for Fuel Oil and Gasoline Shortages

In the spring of 1972 fear was expressed that the U.S. might experience petroleum product shortages by the winter of 1972–73. Obviously, if refineries were not going to have enough crude oil to process there would eventually be shortages of finished products after inventories became depleted. Given the seasonal swings in demand and refinery limitations, fuel oil shortages would likely occur in the winter and gasoline shortages during the summer.

In the early spring of 1972 the OEP received letters and statements from senators, congressmen, and governors expressing concern over the existing and future fuel oil supply situation in the northeast section of the United States. General Lincoln reported to the Oil Policy Committee meeting on April 25, 1972, about the situation.

> This survey, completed April 24, showed conclusively that there had been no shortage, there is no current shortage and none is foreseen in the coming winter [of 1972–73]. Domestic producers indicate that supply is adequate and that additional oil can be produced and delivered in New England if necessary.[11]

In response to the concern expressed over the adequacy of fuel oil supply, General Lincoln wrote the New England Governor's Conference as follows:

> My office has checked with major suppliers on the availability of No.2 heating oil for the coming year and have assurances that if anything, industry capability to produce No.2 has been underused, and industry can supply the heating oil needs for the coming season.[12]

The Empire State Petroleum Association advised the OEP in May 1972 that price controls were working against the building of normal fuel oil inventories. The Office of Emergency Preparedness investigated the situation. The OEP director in Region I found no support for the allegation that price controls were detrimental to stockpiling fuel oil. In Region II another OEP representative came to the same conclusion based upon conversations with Humble, Hess, and Mobil, who indicated that inventories of heating oil should be built as normal for the upcoming heating season.[13]

The National Oil Jobbers Council (NOJC) also wrote to General Lincoln, on May 24, 1972, expressing concern about the fuel oil supply situation. The letter pointed out that:

> Unless prompt, effective action is taken on this matter, many homes, public institutions, industries, and businesses will be without sufficient heat during the oncoming winter. Also, the economic impact upon the independent distributors in our industry will be severe, even to the extent of proving fatal to many smaller concerns.

In June the NOJC Fuel Oil Committee met with General Lincoln and staff to discuss the shortages of fuel oil anticipated by the NOJC for the upcoming winter.

Representatives from the Office of Oil and Gas met with the OEP to discuss the fuel oil situation. These representatives pointed out to General Lincoln that, with the tightness of crude oil, refiners had a choice of what products to emphasize and their decision would depend upon the relative profitability of the product. The choice was basically between producing gasoline or No.2 fuel oil, and the latter's price had been caught by the price freeze near its summer low.

General Lincoln asked whether or not refiners could afford not to produce No.2 fuel oil and he received an affirmative answer in the opinion of the representatives from the Office of Oil and Gas. Elmer Bennett, legal counsel to General Lincoln, indicated that he thought industry was trying to force the government to sanction an industry-wide price increase. General Lincoln replied that he felt a price increase would be hard to justify at that time. Hollis Dole, one of the assistant secretaries for the Department of Interior, said that he planned to talk to major refiners about the matter.

The OEP had a meeting with Humble Oil of Standard Oil of New Jersey (Exxon) to discuss distillate supply and corresponded with Continental Oil Company about the subject. In both instances the petroleum companies indicated to General Lincoln that price would act as an obstacle to meeting the demand for fuel oil.

Officials of the OEP met with representatives of the Humble Oil Company, a division of Exxon, on July 17, 1972. The memorandum prepared for this meeting said that the petroleum industry had the physical capacity to meet demand in a normal winter or a colder than normal winter. Humble pointed out, however, an economic capacity limit was placed on production of No.2 fuel oil since the price of distillate was frozen at the low summer price levels on August 15, 1971. This freeze made it unattractive to produce No.2 fuel oil and worked against providing adequate supplies of this product.

General Lincoln explained that it was unlikely that the ceilings on No.2 fuel oil would be removed and that he expected industry to meet the demand for this product. The memorandum indicated that Randall Meyer then recognized the obligation of industry and stated that industry would endeavor to meet demand. Representatives of Humble said that, "For the 1972–1973 season, Humble will meet demand and provide cold weather protection; they will maintain higher inventories than normal."[14]

Exxon's assurance was not fulfilled; distillate inventories for the last five months of 1972 averaged under 90 percent of the previous year east of the Rocky Mountains. The highest level of distillate inventories in any single month during the period was 93 percent of the previous year. Furthermore, Exxon did not increase their distillate yields to a maximum level until after the first of the year.

The memo also noted that, as of July 1, 1972, distillate inventories were 10 million barrels below the previous year. It was recognized that *"supply could be tight if industry does not operate at required capacity levels."*

Continental Oil Company wrote to the Office of Emergency Preparedness regarding the ability of industry to meet the demand for heating oil during the winter of 1972–73. Like Humble, Continental emphasized that there was adequate physical capacity to meet demand, but that the low, frozen price of distillate fuel oils was acting as a deterrent to the production of necessary quantities. The letter of July 26, 1972, stated that:

It is our present opinion that the industry still has adequate capacity to supply heating oil demand for the coming winter (1972 to 1973), providing the weather is not abnormally cold. However, because of the existing price controls, the industry will have difficulty in making the necessary refinery yield adjustments to supply those demands. If recognition were given to historical normal winter seasonal rise in distillates prices for this coming winter, the petroleum industry would be able to operate at a somewhat higher rate of capacity and thus produce an increased volume of distillates during the winter months. . . .

The matter of price controls appears to be very serious as it affects the overall ability of the industry to maintain an adequate supply of all refined products. . . .

In a memorandum attached to the letter, an OEP staff member noted the conflict between physical capacity and economic capacity. A pencilled note on the memorandum indicated that the industry might be trying to blackmail the government into removing price controls on No.2 fuel oil. The note read:

> I view the above with skepticism. Conoco's rationale doesn't even hang together on logical ground. The Conoco letter is another step in the threat-pressure campaign for price control relief—and nothing more in my opinion.[15]

The pencilled note was signed "E.B."

In an interview on November 12, 1972, Elmer Bennett stated that he and General Lincoln thought that, when the time came, public pressure would force industry to produce. Apparently discounting the likelihood that industry would not produce enough distillate to meet demand and having other pressing problems, the OEP had relatively little contact with the industry during the summer of 1972 about the possible distillate problem.

Second Emergency Increase in the Level of Oil Imports

On September 18, 1972, President Nixon announced the second emergency increase in the level of oil imports. This was a hybrid program, in that the level of oil imports was not simply increased, with refineries being given a one-time chance to import the oil. Instead, refineries were allowed to increase their imports in the final quarter of 1972 by 10 percent of the previously authorized quantity for that year, but this would be counted as a draw against a refinery's 1973 quota level.

At the beginning of September, crude oil and finished product inventories were approximately 60 million barrels below the level of the previous year. Assuming no further change in demand or supply capability, to compensate for this shortage would have required imports during the final quarter of 1972 to have been increased by approximately 650,000 barrels per day. The new Presidential authority would have permitted upward to 600,000 barrels per day of additional imports during the final quarter of 1972 and, had most of this quantity been imported, the petroleum stock situation for that winter would have been greatly improved. Had such quantities been imported, gasoline inventories could have been built to required level in advance of the summer 1973 peak gasoline demand. As was discussed in Chapter 5, only about one-third of the crude oil that refineries were authorized to import was actually used, in a large part due to the way the regulations were written. The urgent problem was to stuff more crude oil into the refinery system to replenish crude oil feed stocks and inventory levels and petroleum product inventories.

A problem with the new import authority was its having come so late in 1972. By grouping such a large increase in the level of oil imports into the final quarter of 1972, it strained the ability of companies to negotiate for additional supplies of crude oil imports and to make economical arrangements for tankers to transport the product to the United States.

General Lincoln briefed the Oil Policy Committee on August 31, 1972, and belatedly noted that the decline in domestic production and the increase in demand for crude oil had transferred the "surge capacity function from the domestic industry to the oil import program. . . ." To restore flexibility to the supply of crude oil, General Lincoln proposed that "all eligible overseas importers . . . be allowed to draw up to 10 percent of their 1972 regular allocation as an advance on 1973 allocation." The chairman went on to "emphasize that any advance would be deducted from the importers' 1973 quota level. . . ."

Lincoln asked that members of the Oil Policy Committee express their opinions of the recommended course of action. According to the memorandum of the meeting, General Lincoln received unanimous, enthusiastic endorsement for his proposed program. Certain people attending the meeting remarked as shown:

> Dr. Solomon [Council of Economic Advisors] commented that the 10 percent idea was an extremely good one which would add needed flexibility to the program.

Mr. Armstrong [State Department] characterized the 10 percent program as "very astute."

Mr. Nehmer [Commerce Department] said that it was "quite imaginative."

Their enthusiastic support for this hybrid and inappropriate program, given the obvious problem of a crude oil shortage starving the U.S. refining industry, simply underscored the naiveté of those responsible for making oil import policies.

Effects of the Borrowing Concept

Three days before the President announced the import program, Ashland Oil Company, the largest of the independent refiners and a supplier of private-brand discount gasoline marketers, attacked the concept which had been reported in the *Oil Daily* on September 13, 1972. Ashland's telegram expressed hope that the *Oil Daily* article was not based upon facts, for such a plan would "squeeze" the independents, since the major oil companies would have no incentive to trade with the crude-deficient independent refineries.

The Ashland telegram recommended that President Nixon should simply authorize an increase in the level of oil imports for the final quarter of 1972, but not adopt the borrowing concept to increase the supply of oil for refining. If the authorized level of imports for 1972 had simply been increased, the international oil companies not wishing to lose the opportunity to profit by additional imports would have increased their imports substantially in the final quarter of 1972. What the federal government would have been doing was "force feeding" the U.S. refining industry with needed crude oil supplies. In essence, this is what Texas and Louisiana had been doing for years through monthly adjustments in the level of production. The regulatory agencies in these two states set the level of state production so that there were normally somewhat higher levels of inventories than the industry wanted. As a consequence the independent refineries purchasing most of their crude oil from others had no problem obtaining needed supplies. The oil import authority seemed not to realize until too late that their responsibility was to see that adequate crude oil supplies were forced into the refining system.

The last chance the oil policy administration had to see that adequate crude oil supplies were made available to the refinery industry was the September adjustment. The fatal mistake at this time was to make the increase in authorized imports as borrowing against the 1973 quota instead of simply increasing the level of oil imports. Whether knowingly or not, the policies of the oil import administration through 1972 created a shortage of crude oil feed stock required by U.S. refineries. Because of the vertically integrated structure of the industry, with some petroleum companies owning and otherwise controlling most of the crude oil needed by their own refineries, the impact of the shortage was thrown largely upon the independent refineries and certain integrated companies that were heavy net purchasers of crude oil. Thus, not only was an unnecssary shortage created, but it also was focused on the independent refineries—a major source of price competition.

On September 7, 1972, eleven days prior to the presidential announcement of the borrowing concept, General Lincoln talked with Frank Ikard, president of the American Petroleum Institute, about proposed changes in the Oil Import Program. This memo was forwarded to Peter Flanigan of the White House. In it General Lincoln stated that Frank Ikard liked the new import program concept—that is, the approach had the blessing of the big oil interests.

A week following the "official" announcement of the hybrid import authority, General Lincoln spoke with Randall Meyer, vice-president of Humble Division of Standard Oil of New Jersey (Exxon). In a *Memorandum for the Record* of September 25, 1972, Lincoln said that Meyer "was certainly glad to see the way we handled additional allocations for 1972."

As it turned out, Humble did not use any of its authority to increase the level of oil imports in the final quarter of 1972 but instead processed crude oil through its controlled refineries that would otherwise have been sold to crude-deficient independent refineries and lesser major oil companies. Had the presidential authority forced Humble to take the opportunity to increase the imports in the final quarter of 1972 or lose the chance for further importation, Humble would likely have had to decide to increase the level of imports and in turn to continue sales of crude oil to traditional crude-deficient refineries. This is just another incident among many where government import policy during 1972 in effect strangled the price competitive independent segment of the industry and favored the large, integrated petroleum companies and crude oil producing companies with strong ties to the Nixon administration. Political patronage was being substituted for proper government policy and in turn feeding the growing petroleum product shortages in the United States.

Summary and Conclusion

Mismanagement of the Oil Import Program during 1972 by the executive branch of the federal government resulted in "starving" the U.S. refining industry of supplies of oil needed to meet demand. Three adjustments were made in the level of oil imports for 1972. Unfortunately, these changes did not result in sufficient increases in the level of oil imports, and they were not made in a timely manner to provide U.S. refineries with oil required for processing.

At the beginning of 1972, the level of oil imports east of the Rocky Mountains was increased by only 100,000 barrels per day— less than 1 percent of the oil processed in the United States. At the time of this decision an OEP staff paper had been circulated indicating that the domestic production could peak in early 1972, and trade journal articles had raised questions about the ability to expand domestic output. Had these warnings been heeded and the level of oil imports increased by 500,000 barrels per day at the beginning of 1972 (it was later increased at the beginning of 1973 by 915,000 barrels per day) rather than by 100,000, the U.S. refining industry would have had ample supplies of crude oil for refining during 1972.

The level of oil imports was increased by Presidential Proclamation on May 11, 1972, by 230,000 barrels per day. In deliberation over the level of oil imports, General Lincoln said that he wanted to increase the level adequately so that another emergency increase would not be necessary later. Unfortunately, General Lincoln did not heed his own advice and ignored the pleadings of independent refineries for increases in the level of oil imports by as much as 500,000 barrels per day and the forecast of one of his own staff economists that the shortage could be as much as 360,000 barrels per day.

On September 18, 1972, President Nixon authorized the final plan to correct the rapidly deteriorating petroleum supply situation confronting the United States. He increased the level of petroleum imports up to 600,000 barrels per day during the final quarter of 1972. With normal lead times of sixty days or more to arrange for and receive shipment of additional oil imports, the new oil import authority came too late to make economical arrangements for additional oil imports. Regrettably, the oil policy makers also erred in their construction of the new import authority, making additional imports a draw against the refiner's 1973 quota. Only about one-third of the additional crude oil and unfinished oils that refineries were permitted to import were actually imported during the final quarter of 1972. Had the additional allocations been made as a one-time, take-it-or-leave-it opportunity, a much larger quantity of oil would have been imported.

It should be recognized in assessing the mismangement of the oil import program that the responsible government officials were receiving mixed advice from different segments of the petroleum industry. While the independent refineries and heavily crude-deficient integrated oil companies were pleading for additional oil imports, several of the integrated petroleum companies and independent producers of crude oil were presenting a counterposition suggesting that further increases in the level of oil imports were not needed. It was to the advantage of many of the integrated petroleum companies and the independent producers of crude oil to have a very tight crude oil supply situation. Crude oil prices were subsequently increased and a crippling squeeze was exerted on the independent refining and discount marketing segments of the petroleum industry.

7

Petroleum Shortages:
An Assessment

As concern for the adequacy of heating oil increased, committees of Congress held hearings late in the summer of 1972 on the adequacy of fuel supplies. Although government witnesses still maintained that the major oil companies were responsible enough to produce sufficient heating oil to meet demand, shortages did occur during the winter of 1972–73.

Much blame was placed on the Price Commission for failing to recognize the need to make a seasonal adjustment in the price of heating oil. Because of a low heating oil price, refiners claimed to have no incentive to produce heating oil. However, the question must be raised whether certain segments of the petroleum industry were using the Price Commission as a scapegoat to cover up their own behavior.

The first part of this chapter will analyze committee hearings on fuel shortages, warnings of various groups concerning the shortage, and the government's realization of and reaction to the shortage. The last section of the chapter is concerned with causes of longer run petroleum product shortages resulting from "inadequacy of refining capacity" and a self-assessment of the performance of government oil policy.

Debate over Likelihood of Shortages

The Senate Committee on Banking, Housing, and Urban Affairs held hearings on September 19 and 20, 1972, on the adequacy of home heating oil supply. The purpose of the hearings was to assess whether or not the U.S. was faced with the likelihood of heating oil shortages during the winter of 1972–73. Several witnesses before the hearing expressed considerable concern over the possibility of a fuel oil shortage. In contrast, government witnesses discounted the likelihood of petroleum product shortages and felt that there were no problems as the U.S. entered the upcoming winter.

General Lincoln, chairman of the Office of Emergency Preparedness, stated at the hearings, "I have been assured by several major suppliers that there should be adequate supply of No.2 oil during the coming winter."[1] General Lincoln went on to state that "the industry has the necessary refining capacity and necessary feed stocks to insure an adequate supply."[2] When questioned by Senator Taft about the contemplated shortage of petroleum products, General Lincoln said, "We don't see a shortage in fuel oil. And we don't see a shortage in other products either."[3]

General Lincoln used the hearing as an opportunity to explain the changes announced by President Nixon the previous day in the Oil Import Program. Lincoln stated that "the President has approved a two-part program which doubly assures an adequate supply of both feed stocks and for No.2 oil. . . ."[4] General Lincoln pointed out that, with the changes made in the Oil Import Program, industry had the capacity and feed stocks "to build stocks to satisfactory levels and to assure adequate supplies of No.2 oil during the coming season."[5] While discussing the levels of refining capacity utilization during the first eight months of 1972, General Lincoln said that industry would have no problem under normal circumstances meeting demand if they would increase capacity utilization to 93 percent in the final quarter of 1972 in the area east of the Rocky Mountains. As was discussed in Chapter 5, during the final quarter of 1972, refining capacity utilization based upon crude oil runs to stills actually averaged only around 89 percent of capacity. With the low inventories of heating oil entering the winter season and the less than all-out production in the final quarter of 1972, there was no way around the shortage of heating oil during the winter of 1972–73.

Steven Wakefield, deputy assistant secretary for energy problems in the Department of Interior, also appeared at the Senate hearings looking into the inadequacy of home heating oil supplies. Wakefield discussed a study that the Office of Oil and Gas had made of refinery operations needed to meet the projected level of demand. This analysis showed that refineries would have to operate at 92 percent of capacity based upon crude oil runs to stills in the final quarter of 1972 and produce a high yield of distillate fuel oil in order to meet the projected level of demand.[6] As to the situation for the upcoming winter, Wakefield said, "We do not see any potential for overall shortages of No.2 fuel oil on a national or on a regional basis."[7]

Senator McIntyre, chairman of the subcommittee, wrote to C. Jackson Grayson, Jr., chairman of the Price Commission, on September 8, 1972, concerning whether or not price controls would keep the petroleum industry from producing the necessary quantities of heating oil. The Senator pointed out:

> The *Wall Street Journal* has recently carried an article in which a vice-president of Shell Oil Company is quoted as saying that without a price increase the refining segment of the industry would not significantly increase its home heating oil refining output.

Mr. Grayson wrote Senator McIntyre on September 14, 1972. Based upon informal discussions initiated with the petroleum industry, Grayson indicated that the petroleum companies did not intend to ask for an increase in the price of home heating oil for the upcoming winter season. He said:

> In view of recent publicity about oil shortages, we initiated informal staff discussions earlier this month with several large oil companies. Those companies have informed us that they have no present intention of requesting a price increase in No.2 home heating oil for the upcoming winter season.

As was previously seen in the conversations between the OEP and Humble Oil Company and Continental Oil Company, price was indeed a problem in the plans of these companies to produce sufficient supplies of No.2 heating oil. It was the industry's responsibility to go before the Price Commission to present their case for a fuel oil price increase if the need existed. The petroleum industry did not do this and did not exert the necessary effort to produce at a sufficient level to meet the demand for fuel oil and the U.S. experienced a fuel oil shortage during the winter of 1972–73.

The Small Business Subcommittee of the House of Representatives held hearings similar to those of the Senate on September 27 and 28, 1972, to look into the adequacy of petroleum supplies for the forthcoming winter. Once again, representatives of the OEP and the Office of Oil and Gas in the Department of Interior testified. Elmer Bennett, chief counsel of the OEP, testified that with the changes in the import program made by President Nixon on September 18, 1972, there would be no shortages of petroleum during the winter. His confidence in the industry's ability and willingness to produce is shown in the following statement. "With an adequate level of crude oil supply assured and adequate refinery capacity, I am sure that the industry will produce the quantities required to meet demand."[8]

The dependence of government on the goodwill of the industry to produce the level of supplies needed by the market place can be seen in the following exchange between Representative Addabbo of the Small Business Subcommittee and Elmer Bennett of the OEP.

ADDABBO: Is the total supply there?

BENNETT: We believe that it is or should be because of the availability of raw material and refineries.

ADDABBO: When you say it "should be," what does that mean—it should be? Either it is there or it is not there.

BENNETT: Well, can we order, for example, refinery A to increase its refinery operations by 5 percent? . . .

ADDABBO: In other words, you are saying refineries by their own choice would rather produce less so that they can lead a false market or a false price-increase situation?

BENNETT: I doubt that very much, but how can I answer the question?[9]

Congressman Hungate of the Small Business Subcommittee questioned Bennett about the new import proposal to permit refineries to draw upon the 1973 import quota up to 10 percent of 1972 imports. Congressman Hungate said that he would have preferred it to have simply been made an increase by 10 percent in the level of oil imports for 1972, instead of being a draw against 1973 import quotas.[10]

Gene Morrell, director of the Office of Oil and Gas, appeared for the Department of Interior at the House hearings. He indicated that he thought the industry would meet the demands of controlled and independent customers for petroleum products. Specifically, he said:

I think that the industry will make enough products, in our opinion, to take care of all the demands that they see at the refinery gate, whether it is an independent marketer or a refiner marketer—I mean, an integrated marketer.[11]

Dave Oliver, economist for the Office of Oil and Gas, stated that he believed the supply of fuel oil was not "as tight as it sounds with regard to 1973."[12] Oliver went on to say: "It is most unlikely there will be deficiency of petroleum supplies in the first quarter or even in the second quarter [of 1973]."[13]

Oliver continued to be optimistic, but wrong, about the petroleum situation. This was unfortunate, for the Department of Interior was a major source of statistical analysis on which the oil policy administration relied.

Supply Situation Worsens and Shortages Become Severe

Despite the optimistic predictions given by government officials in September, the petroleum product situation deteriorated during the fall and into the winter of 1972. U.S. refineries did not operate at 92 percent of capacity during the final quarter of 1972 and demand outstripped production. In addition, the refinery industry did not maximize the yield of distillate fuel oil as the Office of Oil and Gas indicated would be necessary in the final quarter of 1972 to meet demand. The sagging output and the worsening fuel oil situation led General Lincoln to confer with industry and to urge, nudge, and demand higher levels of refinery output and greater production during the final quarter of 1972.

Refinery output in October 1972 fell below that projected by the forecast of the Office of Oil and Gas and that presented at Congressional hearings in September. The worsening petroleum situation led General Lincoln to meet and correspond with several of the major oil companies. Time and time again Lincoln heard from the industry that the culprit was the price freeze on No.2 fuel oil.

A meeting was held with the president and other officers of Humble Oil Company on November 3, 1972. Humble spokesmen stated that refineries would have to operate at 93.9 percent of capacity for the rest of the year to produce adequate supplies of distillate fuel oil, but stated that "economic incentives to operate at these levels does not exist."[14] Among various steps indicated by Humble for relieving the tight supply situation was to "grant price relief by raising the price ceiling" on fuel oil.

The president and other executive officers of Cities Service met with the OEP staff on November 6, 1972. Cities' officials stated that the government could relieve the tight fuel oil situation by ending price controls. Cities' executives pointed out that they were making 1.5¢ more on gasoline than they could from producing distillate and that with the market for all the gasoline that they could produce, there was no incentive for them to shift to distillate production.

Executives of Texaco met with the staff of the OEP on November 8, 1972, to discuss the causes of the distillate heating oil shortage. Texaco affirmed that the frozen price of No.2 fuel oil at its seasonally low level was definitely a factor in the distillate situation. Another problem contributing to the distillate shortage was the difficulty of obtaining adequate supplies of crude oil for the remainder of the year.[15]

Several officers from Sun Oil Company met with the staff of the OEP on November 9, 1972. The representatives stated that the shortage of heating oil was related to the low frozen price of distillate fuel oil. They also pointed out that the low level of distillate heating oil inventories entering the final quarter of 1972 was a major factor contributing to the shortage of fuel oil being experienced. The only real alternative in Sun's opinion was whether there was to be a distillate shortage this winter or a gasoline shortage next summer.[16]

Officials of Shell Oil Company also met with the OEP staff on November 14, 1972. The Shell representatives presented a rather detailed explanation of why refineries had not operated at the forecasted 93 percent of capacity level but instead at around 90 percent during the month of October, the beginning of the heating oil season. Shell stated again and again in their written presentation that the major contributing problem was the price freeze on No.2 fuel oil. They advised the government that enough No.2 fuel oil could be produced to meet demand if price relief was given immediately. In essence, Shell was saying that if the government did not give price relief on No.2 fuel oil, it was very likely to be a long, cold winter.

Shell had prepared an analysis of refining operations that would be needed to meet winter demands for No.2 fuel oil. The Shell report, dated September 5, 1972, stated that for the final quarter of 1972 and the first quarter of 1973 refineries would have to operate at 93 percent of capacity and produce 25 percent distillates to meet forecasted demands for a normal winter. This level of refining operations and distillate yield was based "on the relaxation of 'price freeze constraints on No.2 fuel oil prices. . . .' " Shell went on to indicate that "unfortunately, no price relief has been allowed on No.2 fuel oil and the industry has operated refineries at about 90 percent of capacity rather than the forecasted 93 percent." The consequence of this action was that the petroleum industry produced approximately 155,000 barrels per day less distillate than forecasted over approximately the six full weeks preceding the OEP meeting with Shell.

Shell's data also showed that unless refineries operated at 93 percent of capacity, the winter fuel oil shortage would lead into a gasoline shortage at the beginning of the summer driving season. This is exactly what did occur with refinery runs at about the 90 percent level. Throughout the winter of 1972, gasoline inventories were not built up as normal in advance of the peak demand period for gasoline.

To alleviate the forecasted shortage of No.2 fuel oil to be followed by a gasoline shortage, Shell recommended:

> that immediate action should be taken by the Price Commission to provide industry with sufficient economic incentive to maximize No.2 fuel oil manufacture. If there is a normal winter, there might be sufficient time for the industry to take corrective action in the distillate area provided prompt price relief is forthcoming. . . .
>
> While the precise amount of the incentive is difficult to predict it is our best estimate that an increase of about 1.5¢ per gallon in ceiling prices would be adequate to produce an average refinery utilization level and will also contribute to needed increases in gasoline inventories. We believe that 93 percent is the maximum practical limit for sustained periods of time. . . .[17]

General Lincoln met with L. M. Ream, Jr., executive vice-president of Atlantic Richfield on November 17, 1972, to discuss the basic economics of the petroleum industry. In a follow-up letter dated December 1, 1972, Ream stated that, "we continue to strongly urge near-term price relief, if we are to alleviate a continuing supply crisis of even greater dimension than the present one." Ream warned of the need for tough executive action saying:

> We respectfully submit that the only effective response that can be taken to the situation as we analyze it, is prompt Executive action to afford some relief from government price constraints to meet our mutual objectives of sufficient supplies of heating fuel available at a reasonable cost to the consumer.

OEP Response to Alleged Pricing Problems

In the previous section it was shown that major oil companies were emphasizing the point that the low frozen price of heating oil was acting as a deterrent to the needed production of No.2 fuel oil. However, the Price Commission required public hearings before it would consent to a No.2 fuel oil price increase. In such a hearing the large petroleum companies would be required to reveal to the government and to the public information they preferred to maintain as confidential. Instead of formally requesting a public hearing, the giants of the petroleum industry badgered the OEP to intervene for them to obtain an increase in the price of No.2 fuel oil.

Finally, in January 1973, the price controls on No.2 fuel oil were removed and the price was increased about one cent per gallon. As this occurred, the relative production of distillate heating oil increased at the expense of gasoline production; but this was too late, and too much production had been lost to avoid continuing shortages. Now, when it was too late to prevent shortages from occurring, the industry presented its case for increasing the price of No.2 fuel oil. Why didn't the petroleum industry request such a hearing in the summer or early fall of 1972 when there was still time to avoid a shortage? Was the reason for not requesting an early hearing because a price increase could not be justified at that time as alleged, time was needed to psychologically prepare the stage for forcing through a price increase, or was there some other, less obvious, legitimate reason for not making a formalized request with the proper authorities for an increase in the price of No.2 fuel oil?

"Capabilities are rarely equal to intentions," said Philip Essley, an in-house economist for the OEP, in a review paper that he prepared, dated November 1972, on the *Distillate Fuel Problem*. Essley concluded in his paper that "the industry has the capability to meet anticipated demand, even assuming a colder than normal winter." He went on to warn that unless "measures are taken to induce or force greater refinery production," the anticipated demand will not be met. According to Essley, the problem was the price freeze on fuel oil at a seasonally low level and the resulting lack of economic incentive to produce No.2 fuel oil. Essley cautioned the OEP that "in the absence of a price increase, or some government action to force, or induce, refineries to increase distillate production," distillate production would continue to lag behind demand.

The Joint Board on Fuel Supply and Fuel Transportation was created by President Nixon on September 29, 1970, and was located in the Office of Emergency Preparedness to deal with short-run fuel supply problems. The minutes of the November 1972 meeting noted that the U.S. refinery industry was not operating at production levels required to meet demand, refineries were operating below capacity, and gasoline was being made at the expense of fuel oil. It was acknowledged in the minutes that price controls could be a "disincentive to fuel oil production." The report concluded that, unless refineries were induced to increase production, there would be a shortage of fuel oil during the upcoming winter. The belief was expressed that crude oil supplies were ample as a result of the President's increasing the allowable level of oil imports on September 18, 1972. It was noted that in recent weeks refineries had been operating below capacity and had not devoted as much of their production to distillate as projected in the September analysis. As a result of constrained refinery activities, distillate inventories were not being built as rapidly as needed to meet the peak winter demand for fuel oil.

In a memorandum for the Honorable John Whitaker of November 14, 1972, the action of the Joint Board on winter distillate supply was presented. Whitaker was informed that, while industry had engaged in a great deal of informal discussions on the need for a price increase on No.2 fuel oil, no formal application had been presented to the Price Commission and it was the industry's responsibility to present its own case. A statement to this effect in the memo reads:

> While the industry has engaged in a great deal of informal discussion on the need for a price rise, no formal application to the Price Commission has yet been made. Several companies have strongly implied to the OEP that they hoped some Government agency concerned with the problem will take the initiative. They have been told that the responsibility for making the case rests with the industry. . . .

It was further noted that the government simply can not sit back and "do nothing, but must plead with the industry for increased production of fuel oil." General Lincoln said:

> We need . . . to move quickly on the appeal to industry because of the immediate need to strengthen our distillate situation. Also, if we fail

to act now, we would be subject to the criticism that we did not alert an appeal to the industry on a timely basis.

General Lincoln, on November 16, 1972, in OEP Press Release Number 522, said that "the refinery industry must increase production of distillate fuel oil in order to meet anticipated needs." This decision had been reached by a meeting of the Joint Board on Fuel Supply and Fuel Transportation. The Press Release noted "a need for a general and immediate effort by the refining industry to increase productions of distillate in order to assure the needed supply." As a followup on this press release, on Friday, November 17, 1972, the Department of Interior sent telegrams, along with a copy of Press Release Number 522 which called for increased output of distillate fuel oil, to every refinery east of the Rocky Mountains.

Lincoln addressed the National Petroleum Council, the association of major U.S. refiners, on December 11, 1972. He recognized that "frozen prices [of heating oil] are a disincentive to inventory-building. . . ." Nevertheless, Lincoln declared that refineries had no alternative but to "increase their refinery production and their output particularly of distillate heating fuel oil."

Early in 1972 it was called to General Lincoln's attention that the low frozen price of distillate fuel oil might contribute to a shortage of fuel oil during the peak demand period in the winter of 1972–73. In June and July General Lincoln discussed the distillate fuel oil problem with his staff but discounted the problem, indicating that if a shortage should occur, industry would act responsibly to meet the demand. Furthermore, General Lincoln noted that it was the responsibility of the petroleum companies to go before the Price Commission to make their case for an increase in the level of distillate prices. When it became apparent in November 1972 that distillate fuel oil production was lagging far behind demand and that the supply situation was becoming serious, General Lincoln held discussions with several of the oil companies concerning the lag in distillate production. The low, frozen price of distillate fuel oil was underscored as being the major contributing factor to the growing shortage of distillate fuel oil.

The seriousness of the situation, with the industry not budging in its position of calling for an increase in distillate price before they would increase production, led General Lincoln to try personal contacts with the Price Commission. He telephoned James McLain of the Price Commission to discuss the distillate fuel oil problem. According to a memorandum by General Lincoln dated November 22, McLain claimed that "the oil industry can make a profit on the current price of No.2 oil and they know it." McLain and Lincoln also discussed the possibility of the Price Commission meeting with a dozen or so companies and warning them either to increase the production of heating oil or face exhaustive public hearings from the Price Commission on cost accounting, etc., on the oil industry.

The position of the Price Commission on the No.2 price situation was analyzed by an internal OEP memo dated November 27, 1972. The OEP felt that little would be accomplished in discussions with the large oil companies. Based on an OEP survey, all the companies contacted claimed to be making economic sacrifices in producing distillate fuel oil at near maximum level. According to the memo, only Cities Service's representatives openly admitted that they were deliberately producing less distillate than possible for economic reasons, and Texaco chose not to respond directly to this question posed by the OEP. Under such circumstances, it was felt little good would come of exerting more pressure on the giant petroleum companies.

In expressing concern over the position of the Price Commission, the OEP memorandum noted that:

> The Price Commission also seems not to understand that a bread or automobile shortage is one thing and a No.2 oil shortage quite another.
>
> If the Price Commission embarks on this course of action, walking back at a later date will be extremely unlikely, not to say awkward, and the elapsed time (and production) would be lost even if they were to do so.

The OEP analysis went on to say that some critical questions should be answered very precisely:

1. What are the benefits to be gained from the Price Commission's refusal to raise the No.2 ceiling?

2. How do these benefits compare with the obvious cost associated with a severe No.2 fuel oil shortage?

The OEP memorandum reached the conclusion that the fuel oil shortage was partially related to the stubbornness of the Price Commission about their position. The report noted that the "crisis stems directly from the failure of the Price Control system to recognize and provide for seasonal adjustments of fuel oil prices."

Yet the attempt to place the blame for the fuel oil shortage at the doorstep of the Price Commission was certainly an oversimplification of the situation. As already noted, the oil import policy was poorly managed and contributed to starving the petroleum industry of needed crude oil feed stocks. Under shortage conditions refineries were actually given the choice of stressing one product at the expense of another. With gasoline being more profitable to produce, it was understandable why No.2 heating oil was deemphasized and gasoline stressed.

One must not forget industry's role in the shortage. The giants of the petroleum industry could have imported more crude oil and produced fuel oil had they wanted to even if it had been at a sacrifice. However, some of the large petroleum companies made decisions that were in their best economic interest. As a consequence, the U.S. experienced a fuel oil shortage and then a gasoline shortage. With the petroleum industry being so basic and essential to our economy, was this appropriate and responsible behavior by several of the large refining companies? Don't these companies have a responsibility to do everything within their power to support the public interest, even at a temporary small financial sacrifice to their own short-run interests?

Following the call to the Price Commission and the in-house discussion of the position of the Price Commission regarding the price controls on No.2 fuel oil, General Lincoln called Peter Flanigan, special assistant to President Nixon. According to Lincoln's interpretation of the conversation, Flanigan declared that the pricing question was the responsibility of the Price Commission and that he had confidence in its officials to act appropriately in this situation. The memo by Lincoln noted the following:

We discussed the pricing situation. I explained that this rests with the Price Commission and is being handled by Lou Neeb. He [Flanigan] expressed confidence in Neeb and indicated that perhaps we should wait to see what Neeb comes up with.[18]

In a memorandum to President Nixon dated December 18, 1972, General Lincoln stated that the No.2 oil shortage could be traced "in large part to the pressure on refinery capacity coupled with the price control situation which has made gasoline production preferable to

production of heating oil." He also indicated that another contributing factor to the growing shortage of fuel oil was new emission standards for cars which increased gasoline consumption by 300,000 barrels per day. What General Lincoln did not point out to the President
was that crude oil supply was very tight and that crude oil inventories were very low. This problem, which inhibited refineries from
operating closer to capacity, largely resulted from the shortcomings
of the oil import policy.

In another memorandum to President Nixon, dated December
21, 1972, Lincoln referred to a letter that he had sent on December
18, 1972, to C. Jackson Grayson, chairman of the Price Commission,
in which he pointed out that the fuel oil situation was growing urgent
and requested that the Price Commission grant relief on the price
of No.2 fuel oil to encourage increased output by refineries. In this
letter Lincoln pointed out that the price of "fuel oil was frozen at a
seasonal low in August of 1971." Lincoln went on to disclose that,
largely as a consequence of this fact, "distillate production did not
achieve reasonable expectations during the first part of this heating
season." Lincoln further asserted that increased economic incentives
would certainly encourage industry to exert additional effort to
produce No.2 fuel oil. In review, General Lincoln noted that at the
time of the first ninety-day price freeze it was not envisioned by his
office that the price of fuel oil would continue to be frozen at the low
summer level of 1971.

Approximately a month after writing Grayson, General Lincoln
wrote to George P. Schultz, Cabinet member and chairman of the
Cost of Living Council. In this letter, dated January 16, 1973, Lincoln
noted that on December 18, 1972, he had written "the Price Commission requesting an urgent review of the pricing situation for distillate fuel oil." He pointed out that while no reply was received to the
letter, the President's recent action introducing Phase Three (January 11, 1973) did seem to indicate some relaxation "in the stringent
pricing rules which existed previously for distillate fuel oil." Lincoln
noted that "the refining industry has significant capability for further
production although at least part of the further production is likely
to be at higher cost." He stated that he hoped that under Phase
Three interpretation some price relief would be granted to the industry to stimulate the production of distillate fuel oil.

Failure to Increase Level of Oil Imports

On September 18, 1972, President Nixon, at the recommendation of General Lincoln and the Oil Policy Committee, had announced a hybrid program to increase the level of oil imports. Instead of increasing the level of oil imports in the final quarter of 1972 by 10 percent of crude oil imports authorized for the year 1972, the amendment to the Oil Import Program permitted refiners to draw this 10 percent against 1973 import allocations. In November and December, General Lincoln began to puzzle on why industry had not responded to the opportunity to increase the level of oil imports and to bring in the feed stock that would allow them to run at higher levels of capacity utilization. In a memorandum for the record of the OEP (November 29, 1972), Lincoln noted that it was forecasted that domestic production of crude oil would be 150,000 b/d less than originally estimated during the second half of 1972 and that the 10 percent draw against the 1973 quota was based upon this projection.

A survey prepared by the Department of Interior indicated that only about one-half of the 10 percent borrowing authority would be used during the final quarter of 1972.[19] In reality, as will be seen, only about one-third of the authorized crude oil imports were used during that period.

The Department of Interior apparently awakened to the shortcomings of the September borrowing amendment of the Oil Import Program. At the end of November, the Department of Interior recommended that refineries be given a one-time opportunity to use the 10 percent increase in oil imports by February and that it *not* be counted as borrowing against the 1973 quota. As indicated by its change in position, the Department of Interior finally saw that the shortage resulted to a considerable extent from refineries not having the level of crude oil feed stocks needed for processing. To rectify the error in the oil import policy, the Department of Interior was proposing to force-feed crude oil into the refining systems as Texas and Louisiana had done for years.

Designing the new import authority for the final quarter of 1972 as a draw on 1973 oil imports, the oil policy administrators established a plan which, given industry behavior, resulted in a further constriction in the supply of crude oil. This policy played into the hands of the giants of the petroleum industry that own, have under contract, or control through pipelines, most of the crude oil in the U.S. Under such circumstances, a crude oil shortage would have the greatest impact on independent refiners and the heavily crude-deficient integrated oil companies that directly and indirectly depended upon the giants of the industry for much of their crude oil.

The Office of Oil and Gas prepared a report that indicated only 35 percent of the oil import borrowing authority was utilized by U.S. refineries in the final quarter of 1972 (see Table 5–11). Of the 340,000 barrels per day of unfinished oil imports permitted during the final quarter of 1972, only 120,000 were actually imported. The difference of 220,000 barrels per day was vital for stopping the deterioration of crude oil and finished products' inventories that had fallen to dangerously low levels. While the 220,000 barrels per day of imports in the final quarter of 1972 would not by itself have replenished the inventories of crude oil and finished petroleum products, it would certainly have helped the problem.

Ten of the twenty largest refinery companies used none or very little of the additional authority to import crude oil during the final quarter of 1972. Three of the five domestic-based international oil companies—Standard Oil of New Jersey (Exxon), Gulf, and Standard Oil of California—plus Shell U.S. used very little of the additional authority to increase crude oil imports during the final quarter of 1972. Instead, several of these companies reduced and discontinued sales to independent refineries and other crude-poor integrated oil companies.

Starving U.S. refineries of adequate supplies of crude oil played into the hands of the large integrated oil companies with extensive holdings of their own crude oil and control over considerable independently produced crude oil. Staff memos to General Lincoln had indicated that if there were unusual tightness or even a shortage of crude oil, it would be particularly detrimental to the nonintegrated and poorly integrated segments of the petroleum industry. Since this segment of the industry contains many of the price-oriented marketers, government policy contributing to the shortage of crude oil also greatly weakened price competition in the marketing of gasoline and petroleum products.

The tightness and shortage of crude oil experienced by independent refineries and poorly integrated petroleum companies in the spring and summer of 1972 have already been noted. This situation continued for several of these companies through the final quarter of 1972. Following the mid-November communications sent out by General Lincoln urging refineries to step up their output, several of the independents corresponded with General Lincoln.

Ashland Oil Company, in a letter dated November 24, 1972, wrote:

> We have noted your comment that few companies have found it necessary to use their 10 percent borrowing provision against 1973 quota, recently provided by the Office of Oil and Gas. Apparently Ashland is an exception in that regard also. By the end of this year we will have been forced to avail ourself of this entire 10 percent borrowing available to us, and we'll apparently still be short of requirement.

Clark Oil Company also wrote to General Lincoln, on November 21, 1972, and stated:

> We have read with great concern Press Release No. 522 dated November 16, 1972. . . . There is an inference that refineries have not run at capacity from choice; that there are adequate supplies of crude oil feed stock. Certainly this has not been the case with Clark Oil and Refining nor do we believe that it is the case with many other refineries. We have obtained every barrel of domestic crude oil that we could get in 1972, but we are still experiencing a serious shortage.
>
> Assuming, also, the continuation of a very restrictive policy insofar as the levels of crude oil imports are concerned, we have grave doubts concerning our ability to acquire from other companies through exchanges the necessary import tickets with which to supplement our own import licenses, in order to sustain the 88,000 b/d import level.

United Refining met with the staff of the OEP on November 30, 1972. According to an OEP staff memorandum, United stated that "during the past three weeks three of their five domestic suppliers of crude oil had cancelled supply contracts with the result that limited domestic feed stocks had been cut in half."

Shell Oil Company had briefed the OEP in September and November regarding steps which Shell felt should be taken to avoid the heating oil shortages which occurred during the winter of 1972–73. While the steps Shell recommended were finally adopted, they came too late to avoid the winter fuel oil shortages. Shell said, later:

> Phase Three of the Economic Stabilization Act (effective January 11, 1973) removed the price freeze on No.2 heating oil that had prevailed

since mid-August 1971. This action coupled with the elimination of the Western Hemisphere preference on No.2 heating oil and the suspension of quota controls on No.2 heating oil were steps in the right direction and undoubtedly had made some additional material available. Unfortunately, these programs were implemented too late to provide a remedy for the 1972–73 fuel shortage which has been compounded by a worldwide crude oil shortage. . . .[20]

The National Oil Jobbers Council issued a major attack on the operations of the OEP and the executive branch of government for the way they were handling oil affairs. The NOJC asked some penetrating questions regarding the causes of the shortages. They indicated that, in September, General Lincoln had stated that he expected the industry to increase its output in the final quarter of 1972. By the end of the year it was obvious that the industry had not increased production. Industry explained its behavior by saying that it did not have the pricing flexibility necessary to increase output. On the other hand, General Lincoln said that he expected in the next few weeks for the petroleum industry to show statesmanship and to reach the level of forecasted demand. This demand was not met by the petroleum industry, although the NOJC pointed out that the federal government had the authority, power, and responsibility to determine why industry performed so badly.[21]

On January 10, 1972, General Lincoln appeared before the Committee on Interior and Insular Affairs of the United States Senate. He explained that in September his office had evaluated the situation and at that time they had felt that the situation was not a problem. Lincoln stated that what happened during the next few months was that industry did not produce at the necessary level for making enough heating oil to meet demand.

Senator Abourezk inquired of General Lincoln at the January 10 hearing what the OEP had done to alert the Price Commission to the adverse impacts that the price freeze was having on the production of much needed distillate fuel oil. Lincoln stated that he had called to the Price Commission's attention the situation that existed with distillate fuel oil. In response, Senator Abourezk asked, "and nothing came of that, there was no change in their policy?" General Lincoln's reply was that "you're getting a bit out of my field." General Lincoln certainly knew much more about the pricing question and was not giving a forthright answer to the senator's question about what was occurring with regard to the OEP's effort to obtain some price relief on distillate fuel oil.

On February 1, 1973, the Interior and Insular Affairs Committee of the U.S. Senate held hearings on the fuel oil shortage. Randall Meyer, president of the Exxon Company (U.S.A.), testified at the hearings. Exxon laid the blame primarily on the Price Commission and said that:

> The shortage will continue in the future unless both the government and oil companies take the necessary steps to allow expansion of fuel supplies. Exxon believes its recent increase in heating oil price is one of the steps. It would be incorrect to believe that either domestic or foreign sources will be able to continue to meet the nation's heating oil needs at August 1971 prices.

Oscar Wyatt of Coastal States stated that his company had been forced to substantially reduce its refinery output as a consequence of an inability to buy necessary crude oil feed stocks. He indicated that the federal government had not pursued the proper course of action by seeing that there were supplies of crude oil feed stocks as needed by many U.S. refineries. Putting the blame at the doorstep of the OEP, Wyatt said:

> I don't feel that the real problem with the shortage this year lies with the industry. I think that it lies with the Administration's handling of the Import Program in the short term past.
> So I don't really think that the oil shortage should come as a surprise to the government.

On February 22, 1973, the hearings continued before the Senate Committee on Interior and Insular Affairs. Annon Card, senior vice-president of Texaco, presented his analysis of the reasons for the petroleum product shortages. From Texaco's viewpoint the shortages were largely the consequence of the price freeze and poor administration of the Oil Import Program. Card said:

> Price was another key factor in producing the distillate shortage. Under Phase Two wage and price controls, prices for distillate fuel were frozen at the unreasonably low off-season levels in effect during the summer of 1971.
> In July 1972, Texaco advised the appropriate government authorities that one of the facts complicating the supply situation had been a delay in the allocation of required supplemental crude quotas in a timely manner.

In November 1972, Texaco again requested an increase in the crude oil import quotas, however, no action was taken until December 15, when partial crude import quotas for 1973 were announced, permitting refiners to use such quotas in the last 15 days of 1972.

Alan Ward, director of the Bureau of Competition of the Federal Trade Commission, appeared before the Committee on Banking, Housing, and Urban Affairs of the United States Senate that had held hearings in May 1973 on the petroleum product shortages. Ward indicated that, based upon the studies of his department, the shortages were having and would continue to have a particularly severe and crippling effect upon the nonintegrated sector of the oil industry, which was the primary source of price competition. In contrast, Ward said, as competition was being snuffed out by shortages, the profits of the majors were sharply increasing. Ward stated:

First, the supply shortage is a crisis principally for the nonintegrated sector of the oil industry. Some marketing and distributing companies, large and small, may be forced out of business. In contrast, there are no firm indications that major integrated firms face a business crisis. The first effect of the shortage on several of the majors was a sharp increase in their profits.

Long-Term Refinery Capacity Shortage

Ineffective and inappropriate government policy has also been a major contributing factor to a refining capacity shortage problem. By mid-summer 1973, U.S. refineries were being operated at capacity. As shown in Table 7–1, refinery capacity for the years 1971 to 1975 was, or will be, expanded by approximately 400,000 barrels annually, or only a little over 3 percent per year. Over the same period demand has, or is expected to be, increased by 4 percent to 7 percent per year. The lagging growth in refining capacity is forecasted to cause shortages in 1974 and 1975.

The move toward a shortage of refinery capacity since 1970 has been blamed on three primary problems. One of these was uncertainty over the course and direction of the Oil Import Program. A second factor was the poor rate of return which became increasingly important with the new emphasis of the industry on making downstream profits from refining and marketing investments. The third causal factor has been the environmental developments that

TABLE 7-1

Expansion of Refinery Capacity

Year	Barrels Per Day Capacity Added	Percent Increase in Capacity
1977	1,165,000	
1976	896,000	
1975	1,025,000	a
1974	429,500	
1973	312,300[b]	
1972	419,553	3.2%
1971	376,570	3.0
1970	775,955	6.5

a. The projected figures for 1973 to 1977 do not consider the shutdowns of old and obsolete equipment that will take place.

b. In actual fact, the industry expanded refining capacity by some 833,000 b/cd in 1973, by a process referred to as "debottlenecking." See the refining edition of the *Oil & Gas Journal,* April 8, 1974, p. 8.

Sources: Projected figures—*Oil Daily,* November 9, 1973, p. 17. Actual figures —Annual Survey of the Bureau of the Mines.

have called for significant refinery investment to permit the manufacture of low-lead and unleaded gasoline. A further environmental related problem has been obtaining sites for the construction of new grass roots refineries.

Uncertainty surrounding the future direction of oil import policy seems to have been the most important contributor to the growing shortage of refining capacity. President Nixon announced on April 18, 1973, that the oil import quota system would be replaced with a more flexible fee system. The fee system gave preference to the importation of crude oil over refined products and permitted refineries to import the quantities of crude oil that they would require for processing. With this long overdue change and clarification of oil import policy industry announced over the next six months plans to expand refinery capacity by 2 million barrels per day. Unfortunately, most of this capacity will not become available until late 1975 and 1976 (see Table 7-1) because of the time required to bring-on-stream new capacity. In the interim, the United States will probably find that shortages are a way of life, since refining capacity will be unable to keep up with demand.

The *Oil & Gas Journal* in an article in early 1971 pointed out in a headline, "U.S. May Be On Brink of Refining Capacity Crunch."[22]

By mid-year 1971 General Lincoln recognized that, in the future, refining capacity problems might occur if the government did not take action to stimulate the expansion of refining capacity in the United States. In a memorandum to Presidential Aide Peter Flanigan, June 21, 1971, Lincoln said: "I have been struggling for some time with worry that we may need to bite the bullet soon on an explicit policy to assure refinery capacity in the United States; otherwise, it may be too late."

Unfortunately, the oil policy administrators did not "bite the bullet" soon enough. When the federal government finally did act on the problem in April and May of 1973, it was indeed too late, and the U.S. was set for a shortage of refining capacity needed to meet the projected level of demand in 1974 and 1975.

The prospects for a refinery capacity crunch remained a central topic in the *Oil & Gas Journal*. In its November 15, 1971 issue, the refinery problem was presented in an article entitled "Refining Capacity Now Number One Issue in Washington." The lag in expansion of refinery capacity was discussed and the article stated that General Lincoln "warns against allowing this situation to develop." At the same time, however, the article stated that the refinery problem was related to uncertainty over government oil policy. The situation was summed up as follows:

> Refineries at this point aren't sure which way to jump. A plant built in the U.S. to supply a product which can be freely imported would go bankrupt.
>
> On the other hand, a plant built offshore to supply the U.S. could be a white elephant if U.S. regulations are changed to emphasize crude imports and discourage further increases in import of products.[23]

A year following the warning of future shortages in refining capacity, the *Oil & Gas Journal* once again emphasized this ominous problem. In the March 25, 1972 issue there were two articles entitled "Refinery Trends Are Submerged by Flood of Uncertainty"[24] and "Work Start Needed Soon to Avoid '75 Gasoline Gap."[25] Both of these articles stressed that time was running out to "avert a refining-capacity crunch."

The need to make some decision and to move ahead with the task of expanding refinery capacity was communicated by several of the oil companies to the OEP from July through November 1972.

In a letter July 28, 1972, from Maurice F. Granville, chairman of the board of Texaco, to General Lincoln, Granville stated that, "One of the more obvious factors affecting the proposed new refinery capacity over the next several years is the uncertainty surrounding import policy." On September 13, 1972, General Lincoln wrote Peter Flanigan about a briefing concerning the refinery situation that he had received from officials of the Sun Oil Company. General Lincoln wrote, "If the analysis is correct, and it is persuasive, it shows, for instance, that we are going to be short of refining capacity no matter what we do by 1974." General Lincoln closed his note to Flanigan by stating, "I recommend that you get this briefing in the near future." Cities Service executives met with officials of the OEP on November 6 and indicated that government policy regarding the oil import program, environmental standards, and price controls was acting as a deterrent to the needed expansion of refinery capacity. Finally, Humble Oil Company, in a meeting on November 3 with the OEP staff, stated that the company was "contemplating no additional refineries in the United States" because of "environmental barriers such as land use and siting, and because air and water quality standards would be too difficult to overcome." Similarly, at a November 8 meeting, Texaco pointed out to the OEP that "environmental standards and environmental siting problems" were costly and detrimental to new refinery construction.

Even officials in the executive branch of the federal government expressed frustration over the slight progress that was being made on the growing refinery capacity problem. Hollis Dole, Assistant Secretary of Interior for Mineral Resources, wrote to General Lincoln on September 1, 1972, to discuss the refinery capacity situation. He said:

> We believe that it is imperative that a proposal be designed to encourage the construction and expansion of domestic crude oil refining facilities to assure that this nation shall not become overly dependent upon foreign refining capacity.

The deteriorating refining situation also bothered Steven A. Wakefield, Deputy Assistant Secretary of Department of Interior. Appearing before the Senate Banking Committee on September 21, 1972, he described the nation's refining capacity as being in a "precarious state."

The new director of the Office of Oil and Gas, Duke R. Ligon, prepared a memorandum examining why refining capacity was not being expanded. In this memorandum, dated January 31, 1973, to Deputy Secretary of Treasury William Simon, Ligon noted:

> For the last decade, and especially in the last five years, uncertainty over the structure of the U.S. Mandatory Oil Import Program has inhibited planning of U.S. refineries.
>
> . . . The [Import] program has been administered on an unstable and *ad hoc* basis. New features have been inserted into the program to deal with current crises or to reflect special interests.
>
> . . . There has been little evidence that the program has been administered with defined, long range, national objectives, or with adequate understanding of the U.S. refining industry which the program affects.
>
> Moreover, inequities have been introduced into the program which discourage investment by some sectors of the industry. The result has been a lag in domestic refining capacity construction and a consequent export of U.S. refining capacity.

On April 18, 1973, President Nixon released his new Oil Import Program. This plan substantially changed the nature of the Mandatory Oil Import Program which had been in effect virtually unchanged since 1959. The quantitative controls over the amount of crude oil and finished products that could be imported were removed. Refineries wishing to import oil and petroleum products could do so by paying stipulated fees which varied by product category.

The new import program was designed to stimulate expansion of refining capacity in the United States. The fee structure on finished products was established at a level approximately one cent higher per gallon than imported crude oil. The idea was to encourage the importation of crude oil for refining in the United States, rather than the importation of finished petroleum products. In addition, new refineries were permitted to import duty free 75 percent of the crude oil needed for a period of five years.

With the newly structured oil import program that became effective on May 1, 1973, refineries had a better idea where they stood. Besides, the program was structured to encourage expansion of refining capacity and favored the importation of crude oil over the importation of finished petroleum products. Shortly after the new import program was presented, several oil companies announced plans for substantial refinery expansion and for the construction of new, grass-roots refineries. Unfortunately, the new oil import initiatives came too late to permit the building of needed refining capacity to meet the projected levels of forecasted demand in 1974–75.

Management of Oil Import Policy— and Assessment

The Oil Policy Committee was established by President Nixon on February 20, 1970, and it was to be chaired by General Lincoln, Director of the Office of Emergency Preparedness. This committee was to be composed of the Secretaries of the Treasury, Interior, State, Defense, Commerce, and Justice, and the Chairman of the Council of Economic Advisors. The committee was established to coordinate the actions of departments responsible for oil affairs into a unified, national oil policy. It was responsible for making policy recommendations to the President on the Mandatory Oil Import Program that was established in 1959.

Approximately three years after its founding, the Chairmanship and staff activities of the Oil Policy Committee were transferred to the Treasury Department with the new chairman being William Simon, Deputy Secretary of the Treasury. The transfer of the departmental location of the Oil Policy Committee and the appointment of the new chairman marked a major change in the direction of the administration of oil policy matters.

The internal assessment within the executive branch of government of the past administration of oil policy matters was itself very critical. Dudley Chapman, a representative of the Oil Policy Committee Working Group from the Justice Department and subsequently a member of the White House staff, prepared a memo in early 1973 giving his impression of the oil policy administration. The memo suggested that one of the reasons for shortcomings in oil policy was General Lincoln's refusal to have an executive secretary for the Oil Policy Committee. Such a person would have provided much needed follow-through on topics of importance between the meetings of the OPC and could also have provided a broader based overview tied to longer run objectives.

Chapman was also critical of the way in which the Oil Policy Committee operated. Chapman stated that the Oil Policy Committee

... was initially conceived as an administrative scheme to tap the resources of the affected agencies, and to coordinate policy making in a chairman with minimal staff, [but this] was never realized because of the emphasis on building up the OEP staff.

Mr. Chapman indicated that:

General Lincoln's concept was of a bilateral administration by compromise between the OEP and Interior. The other members were often regarded as sources of problems rather than affirmative sources of input, which he preferred to get from his own staff.

For example, the memorandum indicated that the Justice Department had favored market-oriented reforms of import policies but these had been rejected.

Chapman was also critical of the way in which the agendas and discussion papers were handled for the Oil Policy Committee Working Group and Oil Policy Committee meetings themselves. He said that, "The result is that neither the staff people at Working Group sessions nor the principals at OPC meetings has sufficient time to prepare." The report characterized the Oil Import Program in the following way:

The history of the Oil Import Program is one of *ad hoc* accommodations to crises of the moment. Each one establishes a special exception or other adjustment, which conveys no sense of purpose or principle. The OPC started with the commitment to change this, but quickly fell into the old pattern, from which it has never emerged.

Duke Ligon, Director of the Office of Oil and Gas, prepared a memo dated January 22, 1973, on his evaluation on the shortcomings of oil import policy for the Deputy Secretary of Treasury, William Simon. He said that:

Government oil import policy under the Oil Policy Committee during the last three years has been very unpredictable. It has moved from crisis to crisis, responding to problems and pressures. . . . By taking a more long-range approach, the new leadership of the Committee will have the opportunity to provide a level of certainty to the industry, as to the long-range status and predictability of the Oil Import Program. This should aid in the growth development and stability of the U.S. economy.

In a "talking paper" prepared by Duke Ligon for William Simon, dated February 12, 1973, the following assertions were made:

MOIP (Mandatory Oil Import Program) is supposed to maintain a vigorous domestic, petroleum industry—it has failed.

Domestic production declining, imports rising, refinery construction stalled.

. . . need to rethink what we are doing. Program has operated in a climate of uncertainty, unpredictability, and shifts to meet crisis.
We need:
a better allocation system for import licenses.
a long range approach that is stable and predictable.
a program to stimulate domestic refinery construction.
economic incentives for increased domestic exploration and development of oil and gas.

William Simon, the new chairman of the Oil Policy Committee, held a press conference on April 18, 1973, to discuss modifications in the oil import program. He told the press:

Probably the greatest shortcoming in the present program of oil imports is its uncertainty. As you know, industry can not plan in an uncertain climate. Our import allocations were subject to annual realignment. We were making the estimate. In recent years the program has been altered frequently, and now it is a patchwork of exceptions.

Simon went on to describe the nature of the fundamentally new import program to get oil policy back on a sound basis where the economy would not be experiencing unnecessary shortages of petroleum products.

Summary and Conclusion

Several interrelated factors caused the first peacetime petroleum-product shortages in the United States. These included the low refinery capacity utilization during the first half of 1972 that resulted in a drawing down of petroleum product inventories and created a very tight supply situation for heating oil and gasoline. With the peak winter heating oil demand beginning in the final quarter of 1972, it was necessary for refineries to operate at maximum capacity levels to meet the level of demand. Since refineries did not operate at capacity, it was inevitable that a heating oil shortage would occur during the winter of 1972–73 and a gasoline shortage would occur during the summer of 1973.

The oil policy of the executive branch of government also contributed to the petroleum product shortages. If oil imports had been increased in a timely manner, the petroleum industry would probably have been compelled to operate at higher levels of capacity and much of the shortage could have been avoided. The administration of the Oil Import Program was very short-sighted which resulted in some segments of the refining industry being starved of an adequate level of crude oil feed stock to process into finished products.

Despite assurances received in the fall of 1972 from oil policy administrators that there was no cause for alarm, the United States did experience fuel oil shortages during the winter of 1972–73. When queried about the deteriorating fuel oil situation, the oil companies presented a solid front, declaring that price was a major inhibitor in providing the needed levels of No.2 fuel oil. Finally, after being warned months in advance, General George A. Lincoln, director of the Office of Emergency Preparedness and Chairman of the Oil Policy Committee, acted on the threat that production was being inhibited by the frozen price of distillate fuel oil. Contacts were made by Lincoln with the White House and the Price Commission about the looming fuel oil problem beginning in mid-November 1972, but this was too late to react to the developing situation. The United States was already in the grips of a home heating oil shortage.

Inattention and inaction by oil policy administrators was also largely responsible for the growing shortage of refinery capacity which was forecasted to cause petroleum product shortages during peak periods of demand in 1974 and 1975. In early 1971 the trade journals discussed the worsening refinery capacity problems and by mid-year General Lincoln was on record as being aware of this growing problem. Instead of taking the necessary steps at that time to avert a shortage of refining capacity, governmental officials effectively did nothing until two years later. On May 1, 1973, the Oil Import Program was substantially changed to remove a major obstacle to construction in refining capacity. However, by this time it was too late for it requires several years to substantially increase refining capacity.

8

The Bold Step
to Destroy Competition

The integrated petroleum companies desperately needed to improve their overall profitability as discussed in Chapter 4. Their efforts to improve profitability in the early 1970's (prior to August 15, 1972), however, largely failed, and earnings remained depressed.

One of the real challenges to the integrated oil companies was making downstream refining, distribution, and marketing pay its own way. In the past, when crude oil prices were administered above the competitive level and crude oil profits had been high, downstream subsidization worked to the advantage of the oil companies. However, with growing pressure on crude oil profits, it became necessary to make downstream investments more profitable.

One of the biggest obstacles preventing the large, integrated oil companies from improving their downstream profitability was their outmoded, inefficient, and extremely costly system of selling gasoline to the public. As the major brand petroleum companies increased their gasoline prices hoping to improve their profitability, many of the branded customers changed to the lower priced, private brand gasoline of the independents. In addition, some of the independents that owned their own stations and sold major brand gasoline or major branded dealers who had special rent contracts, started to cut prices, which added to the chaos of the marketplace. When the volume losses became too great, the majors felt they had to reduce their prices to become more competitive. They did this by granting billions of dollars of price subsidies to their branded dealers. Thus, to a very large extent, the depressed profits of the integrated petroleum companies resulted from the major brand marketing system being unable to cope with price competition from private brand marketers and from price cutting major brand mavericks.*

The major brand petroleum companies (e.g., Exxon, Gulf, Standard, Mobil, Phillips, Arco, etc.) could have salvaged their marketing system if management had been willing to close large numbers of unneeded major brand stations. Had this been done, the majors would have lowered their costs, become more competitive, and improved their downstream profitability.

One of the present authors was told by an executive of an oil company with more than 15,000 stations that, for the past two years, his company had seriously discussed closing approximately one-half of their stations. The plan was to concentrate branded volume in the remaining stations and, in the process, become more price competitive and profitable. For several reasons this petroleum company and other majors did not go ahead with sharp reductions in the number of stations. Reasons included aversion to risk, fear of short-term profit loss, uncertainty over reaction of giant competitors, incomplete understanding of problems, some government resistance to eliminating small businesses, and the possibility of solving their marketing probems by increasing their crude oil price so as not to be required to make money on marketing.

With the majors unwilling to streamline their marketing systems in order to become competitive, only one course of action remained for the companies: to short-circuit the competitive process and to use the petroleum product shortages to weaken and destroy the independents and their price method of marketing gasoline to the public. The shortage was an excellent way to tranquilize the workings of competition in the marketplace and to replace it with the might and right of the financially and politically powerful.

---

* A major brand "maverick" is a branded operator that deviated from the accepted norm of behavior of not cutting prices. Normally, "mavericks" were independent marketers or jobbers that owned their own station(s), making it more difficult for the branded system to control their pricing behavior. In certain areas, private branders that had been severely injured by price wars branded their stations to obtain price allowances and became "maverick majors."

Legitimizing the Shortage

To pull off a peacetime petroleum product shortage (that not associated with the embargo), the giants of the industry had to be able to point to external problems. These problems were definitely contributory but not sufficient to cause a shortage, had the large petroleum companies wanted to head it off.

One of the problems practically all of the large petroleum companies pointed to as a cause of the fuel oil shortage during the winter of 1972–73 was price control.[1] Time and again, they instructed General Lincoln and his aides in the Office of Emergency Preparedness that if the price controls were removed on No.2 fuel oil (i.e., home heating oil) enough fuel could be produced to meet demand. Thus, by their own admission, they had the ability to avoid the initial fuel oil shortage, but were not inclined to do so without a price increase.

A study of the communications between the large petroleum companies and the executive branch of the federal government causes one to raise the question of whether the larger integrated petroleum companies actually wanted an increase in fuel oil prices, or primarily desired an excuse for a shortage. As General Lincoln correctly noted, the large petroleum companies did not formally petition the Cost of Living Council for an increase in the price of No.2 heating oil. Instead they conducted an informal campaign for a price increase with the Office of Emergency Preparedness and the Cost of Living Council. It was not until very late in 1972 that two of the smaller petroleum companies, Ashland and Murphy, formally requested increases in the price of No.2 oil. According to procedures of the Cost of Living Council, formal hearings would have to be held to consider whether the desired price increase was justifiable. Hearings were held in early 1973 with the change to Phase IV price controls, and the petroleum companies were permitted to increase their No.2 fuel oil prices. Unfortunately, this price increase did not come soon enough to induce the petroleum industry to build fuel oil inventories prior to the winter or to maximize capacity during the winter.

Another factor the petroleum industry has frequently named as causing the petroleum product shortages was a shortage of refining capacity caused by the environmentalists and environmental problems.[2] It was alleged that environmental organizations prohibited the building of new refineries. Admittedly, there had been protests against the building of some new "grass-roots" refineries, and, in cases where requests were made to build refineries contiguous to resort areas or in bird sanctuaries, there may have been reason to question the selection of such sites. A survey conducted by the Senate Commerce Committee in late 1973 added some light to the environmental role in the refining questions. Each major was asked how many refineries it had specifically proposed in the continental United States since 1958 that were not built because of citizens' law suits based upon environmental issues. Of the ten majors[3] responding, not one had failed to build a refinery because of environmentally based civil law suits.

Russell Train of the Environmental Protection Agency offered a different explanation of why new refineries had not been built. He pointed out that the major reason for refineries not being constructed was the oil import program.[4] His assessment seems to be fairly consistent with the facts. Following the removal of oil import restrictions in April 1973, the petroleum companies announced plans to build several new "grass-roots" refineries and to add significantly to refining capacity.

It should also be noted that, even before environmental questions were raised, most of the new refining capacity in the United States had resulted from "debottlenecking"* and expanding refineries. For such expansion programs, environmental questions were often not even raised. Some idea of the industry's ability to expand refining capacity was revealed in a study by the Department of Interior in 1973. The report indicated that, within a year, almost a million barrels per day of refining capacity, approximately 6 percent of the total, could be added by debottlenecking alone. The facts seem to indicate that, while the environmentalists were certainly around and vocal, the refinery capacity crunch was not largely their doing.

When discussing the refining shortage question, one should remember that the petroleum product shortage of late 1972 and 1973 was not related to inadequacy of refining capacity. At that time, there was sufficient refining capacity had it only been scheduled and used

---

* Debottlenecking means eliminating production restrictions in a process by adding capacity at stages that are restricting the overall output of a system.

properly. In fact, it was the underutilization of refineries in the first part of 1972 that had created the very tight supply of petroleum products entering the final quarter of 1972 (see Table 5–1).

To avoid petroleum product shortages during the winter of 1972–73, refineries had to operate at capacity during the final quarter of 1972 and in the early months of 1973. However, the refining industry did not increase its output to capacity as needed. One of the explanations offered as to why the refining industry did not produce when "under the gun" to expand output was a supposed shortage of sweet crude oil* needed for processing by U.S. refineries.[5] Yet the question must be raised as to whether there really was a shortage of sweet crude oil or whether this was just another excuse for the shortage. For example, when James E. Lee, president of Gulf Oil Corporation, was interviewed by a Philadelphia *Inquirer* reporter about the petroleum product shortages during the first half of 1973, he was quoted as saying, "The problem we have in the United States is not a shortage of crude oil. We can import whatever we want."[6] Furthermore, at the time that the U.S. was experiencing its shortage of petroleum products, the rest of the free world was suffering no recognized shortage of petroleum products. The *Inquirer* report pointed out that Exxon and other international oil companies were advertising for more petroleum customers outside the U.S. in early 1973, while this country was experiencing a shortage of supply.

Somehow, the shortage of crude oil for processing by U.S. refineries was suddenly solved and refining capacity utilization spurted in the second half of May 1973. For June, July, and August, refinery capacity utilization increased three to four percentage points, to around 94 percent of capacity utilization (crude oil runs to stills) east of the Rocky Mountains, yielding around 400,000 more barrels per day than during the final quarter of 1972 and the first four and a half months of 1973. By the time refineries got around to churning out products at capacity levels, however, it was too late. The petroleum product shortage had already become severe and supplies had been abruptly terminated to many private brand, discount gasoline marketers. By mid-year 1973, after the shortage effect had been felt, refinery output increased and the supply situation eased—but not

* Sweet crude oil has a relatively low sulphur content. Most refineries in the United States were constructed to operate on sweet crude oil and would be destroyed by processing "sour" crude oil.

for the independents. The supply valve to the independents had been cut back and when conditions eased in the early summer and fall, it was the majors who were pumping gasoline while the independents continued very short.

Control over Independents' Supplies

A complex set of interrelationships and dependencies exists between vertically integrated companies and independent oil producers, refiners, and marketers. The nature of these relationships created the mechanism by which the giants of the industry could exert a severe supply and price squeeze on independent price marketers and independent refiners. What was needed to initiate the squeeze was the petroleum product shortages as previously described.

Considerable light was shed on the inter-functionings and control mechanisms of the petroleum industry by the Federal Trade Commission preliminary report on its *Investigation of the Petroleum Industry* dated July 1973. To grasp the workings of this complicated industry, the FTC decided to focus its study on two of the five petroleum districts in the country. One of these was the largest producing and refining district (PAD 3), which includes Texas, Louisiana, and other Gulf Coast states. The other area was the largest consuming area of petroleum products, the East Coast of the U.S., from Maine through Florida (PAD 1).

For its analysis, the FTC divided the petroleum industry into five basic segments:

1. Eight largest integrated petroleum companies: Exxon, Texaco, Mobil, Gulf, Socal, American, Shell, Arco;
2. Ten next largest integrated petroleum companies: Union, Phillips, Continental, Cities Service, Getty, Sohio-BP, Amerada Hess, Skelly, and Marathon;
3. Independent producers;
4. Independent refiners;
5. Independent marketers.

The complex interrelationships between these five segments of the petroleum industry for crude oil production, refinery input, gasoline production, and gasoline sales are shown in Figures 8–1 and 8–2.

The eight largest integrated companies dominate the petroleum industry through their control over industry crude oil operations and production of gasoline. Together these eight companies con-

FIGURE 8–1

FIGURE 8–1

Districts 1 and 3 Crude Flows to Refiners, 1971

(Millions of barrels)

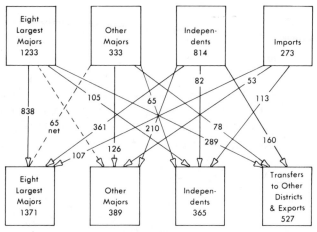

Source: Federal Trade Commission Staff Report, July 1973, p. 11.

FIGURE 8–2

Districts 1 and 3 Gasoline Flows to Marketers, 1971

(Millions of barrels)

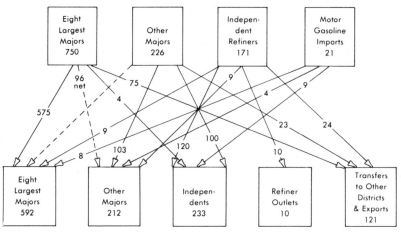

Source: Federal Trade Commission Staff Report, July 1973, p. 11.

trolled 67 percent of the crude oil produced in 1971, having produced 52 percent [1233 ÷ (1233 + 333 + 814)] and bought 15 percent (361 ÷ 2380) from independent producers (see Figure 8–1, top portion). Thus, not only do the eight largest refiners produce a majority of the oil, they extend their dominance by purchasing 44 percent of the independently produced crude oil. In turn, they sell to independent refineries 39 percent [105 ÷ (105 + 65 + 82)] of the domestic crude oil that the independents process (see Figure 8–1, lower portion). As a consequence, they are in a position to withdraw needed supplies from independent refineries, which they did late in 1972 and in 1973, as will be shown later. This ultimately results in a squeeze on independent price marketers since the independent refineries sell them 51 percent (120 ÷ 233) of their output (see Figure 8–2).

The refinery runs shown in the lower portion of Figure 8–1 result in the production of gasoline as seen in the top portion of Figure 8–2. As can be observed, the eight largest petroleum companies refined more gasoline than they sold through their own outlets. In 1971, they sold 91 million barrels of gasoline or 13 percent [(96 + 4 − 9) ÷ (750 − 75)] of their gasoline to other companies. It was this control that they exerted over gasoline which permitted the eight largest integrated petroleum companies to abruptly withdraw gasoline supplies from the independent marketer in late 1972 and early 1973.

The eight largest firms sold less than one percent of their gasoline directly to independent discount gasoline marketers. However, they sold 96 million barrels of gasoline or 14 percent of their output (96 ÷ 675) to other majors ranked nine through eighteen. The other majors in turn sold 100 million barrels of gasoline to the independents. These lesser majors acted as a buffer between the big eight and the independent price marketers.

As long as there was no shortage and the marketplace was directing how gasoline would be sold, the eight largest integrated firms had no real choice but to sell a large and growing proportion of their gasoline to the increasingly more efficient, independent, discount gasoline marketer. It was simply a matter that the largest petroleum companies could not sell the gasoline they refined at the price levels required through their approximately 200,000 branded outlets. Their abundance of stations fragmented their business and drove up the price at which they had to sell. As the majors increased their prices, a growing number of drivers turned to purchasing their gasoline from the discount marketers where they could save several cents per gallon. The large integrated petroleum companies were simply bogged down by the heavy web of their inefficient and backward marketing system, and they were compelled to sell gasoline to the efficient discounters of gasoline. If a shortage occurred, however, gasoline supplies to the independent marketers could be cut, since the eight largest would then be able to force-feed the available gasoline through their own branded outlets.

Withdrawal of Independents' Supplies

As previously discussed, the giants of the petroleum industry were in a position to squeeze independent, discount gasoline marketers and independent refiners, once there was a shortage, by halting inter-refinery sales of gasoline to buffer companies that sold the gasoline to price marketers and by contracting crude oil sales to independent refineries that sold most of their gasoline output to the discounters.

Independent Discount Marketers

The leading edge of the revolutionary pressure on the outmoded major brand marketing system was coming from the rapidly growing, independent, self-service marketers of gasoline. Most of the self service innovators were hit hard by the gasoline shortage that was suddenly sprung upon the marketplace. In addition, many of the more traditional private-brand, discount gasoline marketers experienced severe supply problems. Several examples of the supply problems of self-service operators will illustrate the process of rapid reduction and termination of gasoline sales to many independents.

Autotronics

The volume leader in the field of new companies that entered the self-service business was Autotronics, headquartered in Houston, Texas. Its founder and chairman of the board, James Sadler, identified the bright future for self-service when he had been a vice president of Tenneco. By the spring of 1972, four years after incorporation, Autotronics was the largest of the self-service operators and also one of the biggest independent, private brand marketers. With an organization built and structured for growth, large insurance and bank financing connections, Autotronics had its sights set on a billion gallons per year of gasoline by 1977, at which time it anticipated being the sixteenth largest gasoline marketer in the United States.

Autotronics' amazing growth record and appetite for volume made it a prime target for scuttling by abrupt termination of supply. Data on the company's rapid growth is presented in Table 8–1, where monthly sales volume is expressed relative to the average month of 1972 (e.g., the sales volume in November 1972 was 21 percent higher than the average month for the year and it is expressed as 121.0 percent). In four years Autotronics was operating 550 self-service units and peak sales volume was reached in March 1973. Autotronics' supply was then sharply cut, so that by May 1973 it stood at half the level of three months before. The most dramatic loss of supply came from an almost overnight cutoff from superbroker Armour Oil Company on the West Coast. Another major reduction in supply at about the same time came from superbroker Foremost Oil in the Southwest. Armour Oil's supply in turn had been sharply curtailed by its primary supplier, Gulf Oil, and Foremost's by Arco.

To offset the organizational shock of such a cutback in volume, Autotronics had to make major adjustments in its operations. Hundreds of stations were closed and volume was concentrated in the more profitable units. In addition, executives of the company were terminated and major reductions were made in its organizational staff. The wind was definitely taken out of Autotronics' sails and the company was literally a shadow of the organization it had been.

Autotronics' experience is a classic example of how the independent gasoline marketers fared once the gasoline shortage was set up, triggered, and implemented through the obstruction of the market mechanism. The gasoline supply valve was diverted from the low cost, low price, and rapidly growing self-service type operations to the costly and outmoded major marketing system.

Petrol Stop

Another rapidly growing self-service operation was Timms' and Borgert's Petrol Stop and Petrol Express stations. Starting from thirty units at the beginning of 1968, Petrol stations expanded over a five year period to 273 units primarily operating as self-service

Table 8-1
Sales Growth of Autotronic Systems, Inc.[a]

Year	Jan	Feb	Mar	Apr	May	Jun	Jul	Aug	Sep	Oct	Nov	Dec	Annual	No. Stations
1969														66
1970														239
1971										60.3	70.3	70.2	67.0	391
1972	70.0	72.3	84.6	85.2	92.0	97.6	107.3	109.4	113.6	121.1	121.0	126.0	100.0	550
1973	118.6	116.16	130.7[b]	112.5	92.0	66.8[c]	67.11							

a. Sales expressed relative to average month in 1972
b. High volume month
c. Low volume month following the high

Table 8-2
Sales Growth of Petrol Stop[a]

Year	Jan	Feb	Mar	Apr	May	Jun	Jul	Aug	Sep	Oct	Nov	Dec
1971										74.4	81.2	76.9
1972	85.1	82.6	81.5	92.5	87.2	96.8	105.3	121.3	112.0	112.1	114.1	109.5
1973	122.1[b]	102.4	105.6	67.9	23.2[c]	25.2	37.5	25.0				

a. Sales expressed relative to average month in 1972
b. High volume month
c. Low volume month following the high

stations. The Petrol stations were notorious for their low prices. Most of the units were ex-major brand stations where petroleum companies had decided not to renew leases. Timms and Borgert were able to lease these cast-off stations on very favorable terms, often for only a few hundred dollars per month.

All went well for Petrol until the supply squeeze began in early 1973. Monthly sales in terms of the average for 1972 are shown for the Petrol Stop division that operated primarily in the western part of the country. As can be observed from Table 8–2, sales peaked in January 1973 and fell to one-fifth of that level by May 1973 as Petrol Stop experienced sharp reductions in supply availability. As with Autotronics, Armour Oil Company, furnished by Gulf, had been one of Petrol Stop's largest suppliers. Another significant supplier had been Southern Tanklines, which purchased a large portion of its supplies from Phillips Petroleum Company.

Petrol's operation was the antithesis of Autotronics' approach. While Autotronics grew through construction of new, reasonably attractive facilities, Petrol picked up old and run-down major stations, or, as Borgert referred to them, the major brand "dogs." However, Petrol's method of expansion was just as real a threat to the major brand method of marketing as Autotronics' more capital-intensive approach. If the independents were permitted to pick up major brand "dogs" and reopen them as low price "mini" service stations and self-service stations, they would accelerate the decay of the remaining major brand system. Other companies with less spectacular growth records than Petrol's were doing much the same thing. Examples include Bob Riegel's Texas Discount stations in St. Louis, Gary Simmons' in Kansas City, and Paul Castonquay's Metro"500" in Minneapolis. Interestingly, the Simmons, Metro "500" and Petrol operations no longer exist, very possibly due to the major oil company manipulations. The gasoline shortage was the only way the majors could see to trim down their obese marketing system without its coming back to haunt them.

Winall

The supply difficulties of Winall Petroleum Company are included to illustrate the point that small self-service operators experienced similar difficulties in obtaining supplies as did the larger companies, such as Autotronics and Petrol Express (see Table 8–3).

In 1972, Winall operated eleven relatively high volume, low price self-service stations in the Los Angeles Basin. This smaller, self-service operator experienced severe problems in obtaining supplies as 1973 opened. By mid-summer 1973 its supply position had dwindled to approximately 40 percent of the level obtained in the final quarter of 1972. Under such conditions, Winall, like the other self-service operators experiencing supply cutoffs, was forced to increase its prices, reduce its operating costs, and move from an aggressive marketing posture to one trimmed for survival.

While no major oil companies were directly furnishing Winall, it appears that, based upon records reviewed, they were selling substantial quantities to suppliers of Winall. These included Phillips, the Douglas subsidiary of Continental Oil Company, Arco, and Shell. As the supply situation contracted, as programmed, the majors were able to divert the product away from the aggressive pricing self-service operators to their own costly, major system of marketing that could not survive under normal market conditions.

Pioneer

The three companies discussed, Autotronics, Petrol Stop and Winall, either had all of their operations on the West Coast or a considerable number of stations in that part of the country. Moving now to the South, Pioneer Oil Company is a large self-service operator with extensive market coverage in Texas and parts of the surrounding states.

Pioneer rapidly expanded into a self-service business in the period of 1969 to 1972 as can be observed from Table 8–4. During this period of time, more than two hundred self-service units were opened primarily as tie-ins with convenience grocery stores and other retail outlets. From its peak level of sales volume in December 1972, Pioneer's supply was cut one year later by 58 percent. Furthermore, estimates for January 1974 at mid-month were that supply would be down another 50 percent from the already low level of December 1973.

Romaco

In 1971 and 1972 Romaco, operating primarily in Alabama, made a big plunge into the self-service gasoline business. By early 1973

TABLE 8-3
Sales Growth of Winall Petroleum Company[a]

	Jan	Feb	Mar	Apr	May	Jun	Jul	Aug	Sep	Oct	Nov	Dec
1971												
1972	77.5	75.4	102.9	98.9	104.4	101.3	101.5	102.1	102.8	83.5	74.5	80.6
1973	83.4	86.4	101.6	72.9	65.6	47.1c	47.3	46.0		114.3b	111.7	108.0

a. Sales expressed relative to average month in 1972.
b. High volume month.
c. Low volume month following the high.

TABLE 8-4
Sales Growth of Pioneer Oil Company[a]

Year	Jan	Feb	Mar	Apr	May	Jun	Jul	Aug	Sep	Oct	Nov	Dec	Annual	No. Stations
1969	25.8	25.9	31.0	34.5	38.7	40.2	34.7	36.3	33.2	34.5	37.2	36.4	34.0	109
1970	35.7	33.7	39.0	39.7	44.8	43.6	47.2	51.2	45.6	48.4	52.6	56.3	44.8	152
1971	55.9	52.2	62.1	62.4	67.8	66.2	71.1	74.5	70.7	77.0	76.9	79.9	68.1	231
1972	79.4	68.9	91.3	91.9	95.1	101.4	113.0	108.6	105.9	111.9	114.0	118.6b	100.0	324
1973	101.1	97.9	107.8	97.2	91.8	83.9	87.6	84.0	77.6	83.4	87.5	68.5c	89.0	

a. Sales expressed relative to average month in 1972.
b. High volume month.
c. Low volume month following the high.

Romaco was operating approximately 150 stations and furnishing another 50 units, of which a great majority were self-service outlets. Romaco's pattern of operation was to tie in a one-island, two-pump self-service gasoline unit with convenience stores or other types of retail outlets. From early 1971 to the final quarter of 1972, Romaco's quarterly volume doubled with its push into the low price, low cost, self-service method of selling gasoline.

Romaco's peak period of sales was the final quarter of 1972. Beginning in 1973, Romaco's supplies were sharply reduced as shown below (volume expressed relative to 1972 average):

fourth quarter	1971	76.1%
first quarter	1972	81.1
second quarter	1972	99.2
third quarter	1972	108.7
fourth quarter	1972	110.5
first quarter	1973	100.4
second quarter	1973	47.0
third quarter	1973	40.5
fourth quarter	1973	33.6

From the final quarter of 1972 to the final quarter of 1973, Romaco's supply was reduced by 70 percent. Approximately three-fourths of its gasoline had been purchased from Crown Central. Crown Central was one of the quasi-independent refineries that found the largest integrated petroleum companies ceasing to sell them crude oil and gasoline as they had done for several years. As Crown's supply was contracted, termination notices were sent to many of its unbranded customers. The supply that Crown maintained was consumed largely by its controlled and branded outlets and was not shared equally with other accounts.

Egerton

Lawrence Egerton, a lawyer, had no prior experience in the gasoline business when he opened his first self-service outlet in early 1968. He is properly credited with beginning the wave of self-service in the Carolinas. By the end of 1970, Egerton had opened 110 self-service units. The Olé and U-Filler-Up self-service units were normally tied in with convenience stores and other types of retail establishments. The *National Petroleum News* described Egerton's operations in mid-1970 as "growing like wildfire." The *NPN* went on to state, "He has gained so much success in such a short time that they're [other jobbers] rather awed."[7]

As Egerton and others like him profited by the rapid development of self-service, they took a heavy toll on the costly and outmoded major brand method of marketing. One of the reasons for Arco and BP–Sohio's withdrawal from the Carolinas was the terrible losses in marketing they were experiencing as independent self-service grew.

As can be observed from Table 8–5, peak volume was reached in the second half of 1972 at which time the first cutbacks in supply were experienced. By June 1973, Egerton's supply had been chopped to approximately half the level of six months before and much of the product he continued to receive resulted from legal action brought by Egerton against suppliers. One of the most blatant examples of abrupt termination of supply was by Crown Central, Egerton's principal supplier from the beginning. Crown notified Egerton

TABLE 8–5

Sales Growth of Egerton[a]

1968	Volume	Number of Stations
June 30	2.4%	4
December 31	6.4	11
1969		
June 30	24.4	34
December 31	45.7	67
1970		
June 30	84.4	95
December 31	91.0	110
1971		
June 30	79.6	108
December 31	82.5	114
1972		
June 30	93.5	122
December 31	106.5	124
1973		
June 30	57.8	126

a. Volume expressed relative to 1972 average.

that, effective April 1973, they would no longer be in a position to sell him gasoline.

Egerton had financed much of his expansion through short term bank loans. As supply was cut back and the cost of remaining material greatly increased, Egerton's ability to keep up with bank payments was seriously jeopardized. This was also true of many other independents for whom quick termination of supply meant being forced to the brink of—or into—financial bankruptcy. Their bank payment schedules were based on a level of business operation that suddenly changed. Even the giant oil companies with their huge financial resources and strong financial statements could not have withstood the business interruption and financial jolting that many of the independents experienced. The independents, relying much more heavily on debt, were extremely vulnerable to the abrupt change in their business position.

It has been argued that some of the giant petroleum companies knowingly and deliberately led the rapidly growing self-service operators into a financial trap. Some of the majors were primary suppliers to lesser oil companies and brokers furnishing the independents. All they had to do was to withdraw the supply from the buffer operations and they could whip the carpet from under many of the new, rapidly growing and financially weak self-service operators. If this action did not bankrupt the independents, it would certainly neutralize them so they could no longer threaten the majors in the marketplace.

What the giants of the petroleum industry did was to apply a technique that had been used in other industries at times to clear out the financially weaker independents. It was simply a matter of letting the independents prosper and grow rapidly for a period of time, during which many would burden themselves with debts and other financial obligations. After this period of prosperity would come a rapid change in the business environment that would bankrupt many of the independents and thin their ranks, thus eliminating a significant competitive force in the marketplace.

Marshall

By mid-1972 almost three-fourths of the country's unattended self-service outlets were located in North Carolina.[8] This low-cost, self-service method of marketing by the independents added to the heavy competitive pressures on the hopelessly outmoded, major marketing system. A leader in the field of unattended self-service was Herbert Marshall, who opened 90 coin-operated self-service operations from 1968 to 1972. During this five year period, Marshall's sales volume increased by 130 percent as the public rapidly responded to this extremely low cost method of marketing gasoline.

On June 22, 1972, Marshall was given notice that in two days they would no longer receive supplies from Murphy Oil Company. A scramble for new suppliers led Marshall to obtain a lesser supply of gasoline at a considerably higher price, but at least the company was still in business. As a consequence of the loss of substantial supplies of gasoline, however, Marshall has closed 90 percent of its coin-operated stations, and it is concentrating what volume it has through attended, private brand outlets to preserve its dealer structure. Marshall concluded that it would be much easier to shut down and later reopen its self-service units during the gasoline shortage than to shut down its dealer operation and try to reconstruct that at a later time. If and when Marshall can obtain the needed supplies of gasoline, it plans to reopen its coin-operated self-service outlets.

Independent Refineries

Independent refineries have been one of the principal suppliers of gasoline to the independent discount marketers. For independent refineries to sell to the independents, however, they must have adequate supplies of crude oil to operate their refineries. Several independent refineries experienced withdrawal of crude oil supplies by the giant, integrated petroleum companies during 1972 and early 1973.

The sharp reduction of crude oil sales by Getty Oil Company to several independent refineries during 1972 and the early part of 1973 is shown in Table 8–6. Getty reduced its sales to independent refineries from a high of 29,204 barrels per day during the first quarter of 1971 to a low of 15,386 barrels per day for the first quarter of 1973.

The Senate Commerce Committee sent a questionnaire to the major oil companies asking for information about crude oil sales to independent refineries. A tabulation of the limited responses received is given on page 182.*

---

* The figures are expressed as barrels of crude oil per day and include sales for American, Shell, Sun Oil, Union, Cities Service, Getty, and Standard of Ohio.

Table 8–6
Crude Oil Sales of Getty Oil Company
(in b/d)

	First Quarter 1971	Second Quarter 1971	Third Quarter 1971	Fourth Quarter 1971	First Quarter 1972	Second Quarter 1972	Third Quarter 1972	Fourth Quarter 1972	First Quarter 1973
Sales to Independent Refineries[a]	29,204	25,649	23,017	21,387	14,110	15,632	16,035	15,801	15,386
Sales to Independents as Percent of Total	13.7%	11.9%	11.0%	10.6%	7.0%	7.7%	7.9%	7.7%	7.8%
Sales to Major Oil Companies[b]	184,458	190,221	186,718	180,962	187,160	187,364	186,877	189,785	181,291
Total	213,662	215,870	209,735	202,349	201,270	202,996	202,912	205,586	196,677

a. Independents were American Petrofina, Apco Oil Corporation, Ashland Oil & Refining, Bayou State Oil Corp., Champlin Petroleum, Charter Oil, Clark Oil & Refining, Coastal States, Cross Oil & Refining, Crown Central, Delta, Farmers Union Central, Gladewater Refining, Hunt Oil Co., Koch Oil Co., LaGloria Corp., Lion Oil Co., MacMillan Petroleum, Marion Oil Co., Pennzoil Producing, Permian Corp., Scurlock Oil Co., Tesoro Petroleum, Texas City Refining, Union Texas Petroleum, Vickers Petroleum.
b. Major Companies were Amerada Hess, Amoco Production, Atlantic-Richfield, Chevron Oil Co., Cities Service, Exxon Corp., Gulf Oil Corp., Kerr-McGee, Marathon Oil Co., Mobil Oil Co., Phillips Petroleum Co., Skelly Oil Co., Sohio Petroleum, Sun Oil Co., Tenneco, Texaco, Inc., Union Oil of California.

	1972			1973
1st Quarter	*2nd Quarter*	*3rd Quarter*	*4th Quarter*	*1st Quarter*
126,391	135,607	143,182	136,766	96,910

This data shows that crude oil sales to the independent refineries was sharply reduced in the first quarter of 1973 when the squeeze was tightened on the independent marketers through independent refineries and buffer operations which sold major-produced gasoline to the independents. Unfortunately, most of the larger sellers of crude oil either refused to respond to the questionnaire or avoided answering the question about sales of crude oil to the independent refineries. This included most of the giants such as Exxon, Texaco, Gulf, Mobil, Standard of California, and Arco. Exxon informed the Commerce Committee that the number of independent refineries to whom it sold crude oil had decreased in late 1972 and early 1973, but Exxon refused to reveal the quantities involved. The independents tend to place much of the blame for contraction of crude oil supplies on Exxon and it is unfortunate that government has been unable to obtain comprehensive data on this question from Exxon as well as other large, integrated petroleum companies.

One indication of Exxon's role in sharply reducing sales of crude oil to independent refineries, and in turn cutting the gasoline available to the discount marketers, was Exxon's relationship with Crown Central. This independent refinery obtained 96 percent of its crude oil from other companies, and particularly from the large integrated petroleum companies which controlled more crude oil than they themselves processed. The sharp reduction in sales of crude oil from Exxon Pipeline to Crown Central is shown in Table 8–7. By April and May 1973, Exxon Pipeline had cut its sales to less than one-fourth the average monthly amount for 1972. As Exxon closed the crude oil supply valve to Crown Central, the independent refinery rapidly cut its sales to the independents such as Egerton and Romaco.

Further insight into how the giant petroleum companies used the shortage to squeeze the independent refineries and, in turn, the independent marketers can be obtained by comparing the refinery runs and production of refined products of Crown Central, a small independent refinery with Exxon, the world's largest oil company. As shown in Table 8–8, Crown Central's refinery runs, as a percent of distilling capacity, fell sharply beginning in November 1973 as Exxon withdrew its crude oil sales to Crown Central. Conversely,

TABLE 8–7

Exxon Crude Oil Sales to Crown Central[a]

(Gallons per Month)

1972	Crude Received from Humble (Exxon) Pipeline Company
January	15,127,854
February	13,895,280
March	19,736,976
April	21,548,352
May	33,544,392
June	33,476,982
July	25,778,928
August	22,399,188
September	18,664,002
October	28,878,906
November	23,606,394
December	20,602,134
1973	
January	9,601,620
February	7,756,770
March	9,490,488
April	5,154,618
May	5,647,404

a. *U-Filler-Up, Inc.* vs. *Crown Central Petroleum Corp. and Exxon Corp.*, the U.S. District Court for the Middle District of North Carolina, Greensboro Division, Civil Docket No. C–174–G–73, pp. 11, 12.

Exxon's refinery runs and output turn up sharply starting in December 1972 as they cut back on crude oil sales to the independent refineries. For the first five months of 1973, Crown's refinery output was 25 percent less than for the same period a year before, while Exxon's output increased by 22 percent. (See Table 8–8).

The data also shows how low Exxon runs were in early 1972 as Exxon and certain other companies shorted production in advance of the springing of the shortage on the public in late 1972 and early 1973. With shortage conditions developing, the Nixon Administration cooperated by not increasing oil imports although that could have countered the worsening supply situation. Only under circumstances of shortages could the giants of the petroleum industry actually divert crude oil and gasoline from the more efficient, inde-

TABLE 8-8

Comparison of Refining Operations of Crown Central with Exxon[a]

A. Refinery Runs Relative to Distilling Capacity

	1972												1973				
	Jan	Feb	Mar	Apr	May	Jun	Jul	Aug	Sep	Oct	Nov	Dec	Jan	Feb	Mar	Apr	May
Crown Central	96.0	95.6	94.0	94.7	95.8	97.4	94.3	94.8	94.3	93.1	84.7	73.6	68.8	65.9	69.1	69.2	82.0
Exxon	85.2	85.1	87.5	91.0	93.4	94.9	94.2	96.7	96.6	96.1	96.2	97.7	101.4	101.4	101.5	102.8	103.3
Difference	10.8	10.5	6.5	3.7	2.4	2.5	.1	(1.9)	(2.3)	(3.0)	(11.5)	(24.1)	(32.6)	(35.9)	(32.4)	(33.6)	(21.3)

B. Refinery Output (thousands of barrels per day)

	1972												1973				
	Jan	Feb	Mar	Apr	May	Jun	Jul	Aug	Sep	Oct	Nov	Dec	Jan	Feb	Mar	Apr	May
Crown Central	86.7	89.0	90.2	90.9	91.9	93.5	90.5	91.0	90.5	89.3	81.3	70.7	66.0	63.3	66.3	66.4	78.7
Exxon	830.7	869.7	894.7	930.3	955.3	969.7	964.0	990.0	990.3	983.7	984.7	1014.0	1069.0	1085.3	1086.3	1100.0	1105.0

a. Figures expressed as three-month moving averages.
Source: Texas Railroad Commission for area east of Rocky Mountains.

pendent segment of the combined refining and marketing end of the petroleum business.

Some idea of the extent to which the shortage of crude oil was thrown into the independent refining sector of the refining business was presented by a March 1973 survey of the Independent Refiners Association of America (IRAA). This survey polled independent refineries located in the major U.S. refining area east of the Rocky Mountains. These refineries operated approximately 10 percent of the refining capacity in this area. According to the statement of the IRAA on July 11, 1973, before the House Interstate and Foreign Commerce Committee, the independent refineries surveyed "had a combined operating rate of 1,110,326 barrels per day in 1973 and a combined deficit of 315,734 barrels per day or around 30 percent." Largest quantitative deficits were reported by Coastal States (46,500 b/d out of 135,000 b/d desired rate), American Petrofina (17,810 b/d out of 112,000 b/d), Murphy (15,000 b/d out of 101,000 b/d) and Crown Central (40,000 b/d out of 96,000 b/d).

At the same time the independent refineries were so short of feedstock and had considerable idle capacity, many of the major oil companies had refinery runs of 95 to 105 percent of capacity. The shortage was definitely used to squeeze the independent refineries and in turn their major customer, the independent discount gasoline marketers.

Wholesale Price Increase

There were two basic ways the large, integrated petroleum companies could destroy the independent discount gasoline marketers as a significant competitive force, so that the giants of the petroleum industry could increase their prices and profits. One of the most obvious ways has already been discussed: to sharply reduce the supply of gasoline to the independents. The second technique used to reduce the competitiveness of the independents was to sharply increase the price at which they purchased gasoline.

As 1973 wore on and 1974 opened, the independents' problem was not just a matter of price increases, because the wholesale branded price was increasing as well as the independents' price. The problem was that often independents experienced price increases at a rate of two to three times that of their major brand competitors.

In many cases, the price they paid was near the retail price of their major brand competitors, and allowed no margin to cover operating costs, to say nothing about a fair profit. For the period 1970 through mid-year 1972, the discount marketers had purchased unbranded gasoline generally for 13 to 15 cents less than the normal branded price of regular gasoline at the service station level. When all marketing costs and a competitive rate of return on investment was included, it cost the majors an estimated 13 to 15 cents to market their branded gasoline.

Autotronics provides an example of how the price of gasoline to the independents sharply increased relative to the price of gasoline to its major brand competition. The cost per gallon of unbranded gasoline to Autotronics is compared to the cost of branded gasoline to the major brand jobber. It can be observed from Table 8–9 that Autotronics' price of gasoline averaged about one cent per gallon less than the branded jobber in the period from July 1972 through January 1973. This was very close to what many of the other large private branders experienced in terms of their buying price relative to that of the branded jobber. Part of their difference was accounted for because of the lower actual cost to the oil companies of unbranded gasoline—no advertising and credit card costs. Now, note what happened to Autotronics' buying price as the shortage became pronounced early in 1973. Autotronics' retail buying price increased to, and then above, the branded jobber price. In the second half of 1972, the price of unbranded gasoline to Autotronics was two cents per gallon higher than that of branded gasoline which was more expensive to sell.

Unquestionably, Autotronics' competitiveness has been reduced with the change of circumstances that finds Autotronics paying a considerably higher price for unbranded gasoline relative to the wholesale price of branded gasoline. However, many of the smaller independents have been compelled to pay much higher prices for gasoline. In several instances the wholesale cost of gasoline to the smaller independent marketers has been increased so that it was within one or two cents or equal to the major brand retail price. The consequence of this price escalation to the unbranded, independent marketers has been that they no longer sell gasoline for discounts of three to five cents per gallon as they had before the shortage. Today, it is not uncommon to find the independents selling at major brand

TABLE 8–9
Cost of Regular Gasoline
(In cents per gallon)

Date	A. Branded dealer cost per gallon of regular gasoline (dealer tankwagon)	B. Estimated branded jobber margin for regular gasoline	C. Branded jobber costs of gasoline (A − B)	D. Autotronics' cost	E. Difference between branded jobbers and Autotronics (C − D)
July 1972	17.71	3.50	14.21	13.22	.99
August 1972	17.31	3.50	13.81	13.38	.43
September 1972	18.92	3.50	15.42	13.40	2.02
October 1972	18.47	3.50	14.97	13.48	1.49
November 1972	18.13	3.50	14.63	13.72	.91
December 1972	18.30	3.50	14.80	14.01	.79
January 1973	18.46	3.50	14.96	14.23	.73
February 1973	18.09	3.50	14.59	14.72	−.13
March 1973	18.75	3.50	15.25	15.38	−.13
April 1973	19.02	3.50	15.52	16.48	−.96
May 1973	19.21	3.50	15.71	17.11	−1.40
June 1973	19.22	3.50	15.72	18.17	−2.45
July 1973	19.22	3.50	15.72	18.23	−2.51
August 1973	19.11	3.50	15.61	18.04	−2.43
September 1973	19.13	3.50	15.63	17.98	−2.35
October 1973	20.17	3.50	16.67	18.56	−1.89
November 1973	20.90	3.50	17.40	19.43	−2.03
December 1973	22.53	3.50	19.03	21.26	−2.23

Source: Statement of Ken Catmull, Vice President of Marketing, Autotronics Systems, before Special Subcommittee on Integrated Oil Operations, Committee of Interior and Insular Affairs, U.S. Senate, December 5, 1973.

prices, or at a premium relative to the major brand prices. For example, the *Platt's Oilgram* of January 23, 1974, reported that in Dallas one of the large Exxon car centers was selling regular gasoline at 36.3 cents per gallon on a full service basis with other Exxon stations at 37.6 cents per gallon. The report stated that "these postings are generally below most private brands in Dallas, with the majority [private brands] from 37.9 to 41.9 cents/gallon at self-service pumps."[9] *Platt's Oilgram* of April 1, 1974 stated that "Independents have the product, but they're like the fur coat salesman in the ghetto: the price isn't right." Other journals such as *National Petroleum News Bulletin* also carried accounts showing how the price—and supply—squeeze was being exerted on the independents and threatening their existence.

The price squeeze on the independents is another aspect of the manipulation of the shortage to squeeze and destroy the price aggressive, independent sector of the petroleum industry. The giants of petroleum companies permitted a limited supply of gasoline to flow to the independent segment of the industry, but at the highest possible price that could be finagled. During the summer and fall of 1973, gasoline imports were selling far above the prevailing wholesale price of gasoline and often within a cent or two of the retail price (less taxes). It was this high priced imported product that was often being offered to the independents as a replacement for lower cost domestic supply that had been terminated.

One midwestern marketer, for example, at the end of 1972 was purchasing gasoline for less than thirteen cents per gallon from a petroleum trading company. The trading company had purchased much of the product from the major oil companies, who were unable to sell it through their own high cost and uncompetitive outlets. Less than one year later, that same petroleum trading company was selling gasoline at twice the price, and it was primarily relying on the extremely high priced gasoline imports to furnish its customers. With the shortage, the large petroleum companies channeled the lower cost, domestic gasoline through their own outlets.[10]

The high priced imports of gasoline were only one reason for the rapidly escalating price of gasoline to the independent marketers. A second cause of the spiralling costs of gasoline was that the independent refineries found that they were paying much higher prices for their crude oil feedstock relative to several of their large, integrated competitors. The higher cost raw materials processed by the independent refineries appeared in part to be associated with the integrated oil companies' selling their higher priced imported and domestic crude oil to the independent refineries.[11]

Independents' Loss—Majors' Gain

It was just shown that the petroleum product shortage was used to abruptly cut the sales of gasoline to the efficient, low price, independent gasoline marketers. If the more efficient independents could be forced out of business, their gasoline volume could be force-fed into the major brand marketing system where there was an over-abundance of costly physical facilities to sell the gasoline. Unfortunately, data on the shifts in gasoline volume is rather sketchy. As an illustration of the trend, the Lundberg Survey for October 1973 reported that major brand dealers sold 6.92 percent more gasoline than in October 1972, and that the non-majors sold 3.89 percent less.[12]

Somewhat more detailed data was obtained for shifts in market share of the eight or nine largest petroleum companies compared with "all others" in the state of California. The difference in the relative gains or losses for the eight or nine largest petroleum companies in comparison to "all others" can be observed from the bottom line of Table 8–10. The "all others" category is dominated by private brand sales, particularly those of the independents. What the data shows is that throughout most of 1972, "all others" sales were growing faster than sales of the eight largest petroleum companies. A dramatic shift in the direction of this pattern occurred, beginning in March 1973. From that time forward, the majors' sales improved considerably over what they had been the year before. In contrast, the "all others" category showed a net decline in sales for the same period. This was a consequence of a deliberate program to reduce the supplies going to the independent discount gasoline marketers and to redirect gasoline sales through the major brand service stations.

An anecdote on the supply shortage and cutoff to the independents provides some light on what was happening. One of the large self-service operators experiencing serious supply problems made up some of his lost gallonage from a strange source. On a highly secretive basis, he arranged to purchase several hundred thousand gallons of gasoline from an Exxon jobber who had been given addi-

TABLE 8-10

Gasoline Sold in California as Reported by the State Board of Equalization

1972

	Jan	Feb	Mar	Apr	May	Jun	Jul	Aug	Sep	Oct	Nov	Dec
Standard Oil of California	+7.1	+14.1	+16.1	+13.3	+18.8	+9.2	+1.9	+8.3	+5.0	+5.0	+8.2	+12.5
Shell Oil Company	+4.7	−0.8	+1.2	−9.3	+3.3	−3.9	−4.4	+0.9	+4.6	+7.2	+4.9	+5.8
Atlantic–Richfield	+20.7	+14.6	+9.8	+0.2	−2.0	−0.9	+0.2	−0.5	−6.4	−6.8	−3.1	+0.1
Union Oil Corporation	−9.6	−7.4	−7.1	−11.7	−3.1	−9.8	−3.6	+12.2	0.0	−4.1	+1.7	+5.0
Mobil Oil Corporation	−4.2	+4.3	+9.6	+10.1	+4.3	+11.6	+6.0	+11.1	+6.2	+4.2	−1.4	−0.6
Texaco, Inc.	+28.9	+32.8	+29.3	+12.6	+9.1	+7.7	+3.4	+4.8	−1.5	−2.4	−1.2	−1.8
Phillips Petroleum Company	+1.9	+2.8	+0.7	−13.8	+5.5	+20.5	+11.8	+6.5	−2.0	+3.5	+11.5	+23.4
Gulf Oil Corporation	+3.9	+12.8	−0.9	+2.8	+17.4	+16.0	+1.8	−0.9	+8.6	+12.8	+3.2	−9.1
Exxon												
Eight–Nine Largest	+6.7	+8.3	+7.8	+0.8	+6.0	+4.1	+1.0	+5.4	+6.6	+2.1	+2.9	+4.8
All Others	+12.1	+15.1	+10.4	+3.7	+8.5	+15.0	+3.6	+9.3	−0.1	+20.5	+5.9	+11.6
Difference	−5.5	−6.8	−2.6	−2.9	−2.5	−10.9	−2.6	−3.9	+6.5	−18.4	−3.0	−6.8

1973

	Jan	Feb	Mar	Apr	May	Jun	Jul	Aug	Sep
Standard Oil of California	+6.2	+3.8	+7.2	+8.0	+4.4	−2.1	+4.6	+1.6	+6.7
Shell Oil Company	+8.6	+3.3	+8.6	+16.0	+26.2	+16.3	+11.7	+9.0	+8.0
Atlantic–Richfield	−8.3	−8.7	+4.2	+16.3	−3.8	−4.0	+2.8	+7.3	+11.6
Union Oil Corporation	+6.8	+10.5	+22.1	+24.7	+4.4	+11.5	+20.9	−0.6	+4.1
Mobil Oil Corporation	+2.1	−10.0	−1.6	−1.7	+12.4	−6.3	−2.6	+7.3	+1.3
Texaco, Inc.	−4.3	−10.8	+0.6	+8.6	+10.1	+8.7	+18.9	+15.2	+1.9
Phillips Petroleum Company	+25.6	+11.3	+8.9	+23.2	+4.0	−2.5	+3.8	+4.6	+4.9
Gulf Oil Corporation	+31.6	−0.3	+14.3	+0.7	+9.9	+0.2	+20.6	+15.2	+11.3
Exxon					+38.3	+0.5	+19.7	+6.1	+21.5
Eight–Nine Largest	+5.7	−0.5	+7.5	+11.7	+11.6	+3.1	+9.6	+6.7	+7.0
All Others	+2.3	+2.9	−8.0	−13.2	−18.3	−10.4	−6.8	−11.0	−20.8
Difference	+3.4	−3.4	+15.5	+24.9	+30.9	+13.5	+16.4	+17.7	+27.8

tional quantities of gasoline to sell since the supply shortage had occurred. This report from a normally reliable source of information seemed very likely, especially when one considers that Exxon sales increased according to the Lundberg Survey in October 1973 to 746.8 million gallons—up 100 million gallons, or more than 15 percent, compared to the same month one year before.[13] The same pattern holds true for other reports of Exxon making dramatic gains while the independents suffer crippling losses. For example, *U.S. Oil Week* reported Exxon's tax sales in New Jersey increased by 24.8 percent for the first ten months of 1973 over the same period in 1972, and that October gallonage increased by 55 percent over the same month a year before. Enormous tax gallonage gains were also reported for Exxon, of 55 percent in Connecticut and 59 percent in Florida in October 1973, in comparison to the same period the year before.[14]

A primary contributor to the depressed profits of the major oil companies in the early 1970's (prior to August 15, 1972) was the increased gasoline price wars during this period of time. Harry Bridges, president of Shell Oil Company, noted in a statement concerning "The profit slump: what caused it . . ." that the industry had gone through "the most intense petroleum product price wars of the decade."[15]

One of the basic reasons for the incessant gasoline price wars was the assignment to marketing of a task which it was poorly designed to accomplish—to be a profit center and earn a more satisfactory return on investment. For decades market investment had been subsidized by crude oil profits and not expected to stand on its own. As this occurred the majors' marketing systems were greatly overexpanded and effectively yielded marginal or no rate of return on investment. It was almost an economic impossibility for the vice presidents of marketing of the major oil companies to improve profits by increasing prices, as they endeavored to do time and again in the early 1970's. As the majors increased their gasoline prices, they drove more and more of their customers to the independent price marketers, where discounts were now greater than ever.

As this volume drain continued, the majors felt compelled to lower their prices to become more competitive. As a consequence, price wars accelerated and cost the petroleum companies billions of dollars. The strategy of the major oil companies of increasing their gasoline prices and in turn their profits backfired. What the market-place was signaling to the petroleum companies was the need to streamline their marketing approach if they were going to become competitive and make a profit from marketing.

The frequent gasoline price wars before the petroleum product shortages can be observed from the market data presented in Figures 8–3 to 8–14. The first six charts are based upon weekly surveys of major and private brand gasoline prices for six large, West Coast market areas including Los Angeles, Portland, Seattle, Boise, Phoenix, and Nevada. The next six figures present data on the prevailing major brand gasoline price for six cities located east of the Rocky Mountains. During 1971 and much of 1972, gasoline prices gyrated wildly, falling from "normal" levels to low price war levels.

FIGURE 8–3
Average Weekly Price of Regular Gasoline for Majors and Independents for Los Angeles

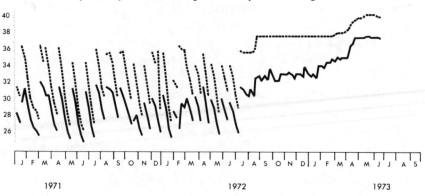

Source: Lundberg Survey.

In the western part of the country, price wars abruptly halted around August 15, 1972, as can be observed from Figures 8–3 to 8–8. It is curious how the giant competitors could reach such market stability so quickly in many markets that had been ripped by price wars for years. To many independent discount marketers, it seemed unreal for the majors to ignore their much lower prices. Over several years the majors had engaged the independents in price wars with differentials of two to three cents per gallon on regular gasoline.

FIGURE 8–4

Average Weekly Price of Regular Gasoline
for Majors and Independents for Portland

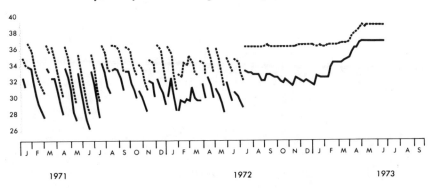

Source: Lundberg Survey.

However, by September and October the independent prices were
frequently five to seven cents lower than the majors. The reason why
the majors did not respond to the discounters' greater differentials
became clear over the next four to eight months, as one price marketer after another experienced sharp cuts in the supply of gasoline
available to him. Eventually the price marketers realized that the
giants of the petroleum industry had not forgotten them, but, on the

FIGURE 8–5

Average Weekly Price of Regular Gasoline
for Majors and Independents for Seattle

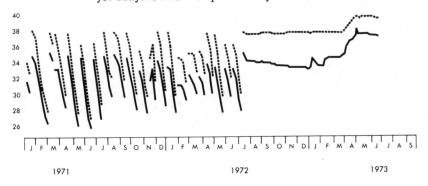

Source: Lundberg Survey.

FIGURE 8–6

*Average Weekly Price of Regular Gasoline
for Majors and Independents for Phoenix*

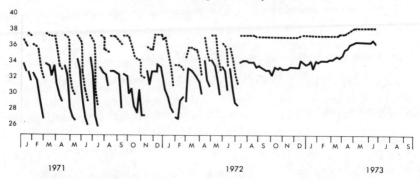

Source: Lundberg Survey.

contrary, were developing the new tool of supply control to throttle the price marketers.

The price charts for Buffalo, Charlotte (North Carolina), and Jacksonville (Florida) show a pattern similar to that observed on the West Coast, where, beginning around August 15, 1972, the majors ceased the practice of extending price allowances to their dealers. When they did this, major brand prices suddenly firmed to a stability seldom seen in the marketplace. The last three price charts for Milwaukee, Miami, and Kansas City (see Figures 8–9 to 8–14) illustrate

FIGURE 8–7

*Average Weekly Price of Regular Gasoline
for Majors and Independents for Nevada*

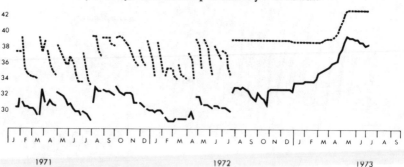

Source: Lundberg Survey.

FIGURE 8–8

*Average Weekly Price of Regular Gasoline
for Majors and Independents for Boise*

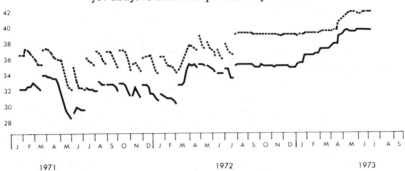

Source: Lundberg Survey.

FIGURE 8–9

*Average Weekly Price of Regular Gasoline
for Major Outlets in Buffalo*

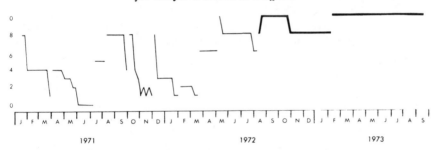

Source: The Oil & Gas Journal.

FIGURE 8–10

*Average Weekly Price of Regular Gasoline
for Major Outlets in Charlotte*

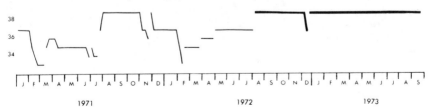

Source: The Oil & Gas Journal.

FIGURE 8–11

Average Weekly Price of Regular Gasoline
for Major Outlets in Jacksonville

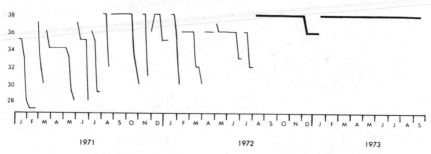

Source: *The Oil & Gas Journal.*

the same effort to bring prices back to normal. However, in some of these markets where independents were still able to get supplies and were relatively strong, the price wars continued for a few months longer. They were finally halted as the cutbacks to the independents began to be felt and destroyed their ability to compete.

The August 15, 1972, nationwide wholesale price restoration can be observed from *The Oil Daily's* Hundred City Survey of the selling price of gasoline to branded dealers (the dealer tankwagon price). About August 15 the giants of the petroleum industry moved in a coordinated effort to withdraw their price allowance programs

FIGURE 8–12

Average Weekly Price of Regular Gasoline
for Major Outlets in Milwaukee

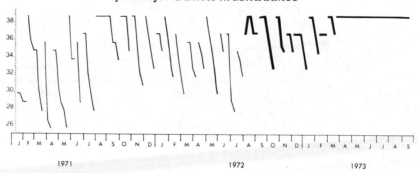

Source: *The Oil & Gas Journal.*

FIGURE 8–13

*Average Weekly Price of Regular Gasoline
for Major Outlets in Miami*

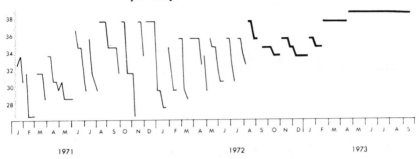

Source: *The Oil & Gas Journal.*

which were used to pay for price wars. In those areas where supplies were under tighter rein and the independents were not as much of a threat, as on the West Coast, the market firmed overnight to an uncanny peace that had seldom been seen before in the gasoline industry. However, in areas of the mid-continent where the independents were strong and still able to obtain supplies, price wars continued until the early part of 1973, as majors struggled to reclaim their gallonage. By March 1973 the giant oil companies had generally with-

FIGURE 8–14

*Average Weekly Price of Regular Gasoline
for Major Outlets in Kansas City*

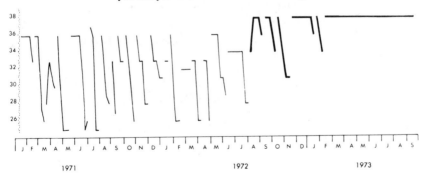

Source: *The Oil & Gas Journal.*

FIGURE 8–15

100 Cities Regular Gasoline Tankwagon Prices

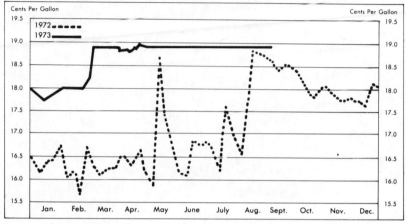

Source: The Oil Daily.

drawn price protection across the country and the dealer tankwagon price stabilized at the price-freeze level.

The August 15 date is particularly worthy of note, for this was the first-year anniversary of the price freeze. This became the day that the giants of the petroleum industry keyed upon to call an end to price wars so that they could rebuild their profit structure. Somehow, rumors were spread through the gasoline industry prior to August 15, 1972, that there would be a new price freeze. If the price marketers and major brand dealers were caught with the prices down, they might have to hold those depressed prices for an indefinite period of time. The major oil companies then led a national price restoration that brought prices back to normal, and they did not restore price war subsidies in many market areas as they had time and again in the past.[16]

Government Delays

The supply squeeze beginning in late 1972 and early 1973 was seriously threatening the existence of many businesses. To dramatize their situation to the Congress and the public, members of the Society of Independent Gasoline Marketers of America (SIGMA)—the largest trade association representing the discount gasoline marketers—met in Washington, D.C., on March 1, 1973. They argued that without government intervention the independent marketing sector would be severely weakened or destroyed, to the detriment of the public. The discount marketers pleaded with the federal government to take action to insure that the shortage would not be used to destroy them. They recommended that their historic suppliers be required to continue to sell them gasoline in the same proportion of the refineries' output that they had purchased the year prior to the shortage (first quarter 1971–third quarter 1972).

Senator Thomas Eagleton's amendment to the Economic Stabilization Act in March 1973 authorized the President to allocate petroleum products as needed to maintain and promote competition and to insure adequate supplies for essential needs. It was envisioned by Senator Eagleton, other members of Congress and by the independents that the President would use this authority to prevent the destruction of the independents. Yet two months following the passing of the Eagleton Amendment, the independent marketers were still pressing the administration to use this allocation authority to insure an equitable sharing of the available supplies.

Administration spokesmen acknowledged the independents' plight but did not require the refineries to share the shortage. William E. Simon, then Chairman of the Oil Policy Committee, said he would not hesitate to use mandatory allocation to prevent undue concentration of hardship on any segment, but he added that he preferred to work through voluntary measures. According to Simon, "It is better to encourage oil companies to move in a desired direction by their own choice rather than to try to force them to move against their will."[17]

On May 11, the Department of the Interior issued a news release stating that the Office of Oil and Gas would immediately begin to administer a voluntary program for allocating crude oil and refinery products. A month later, the voluntary allocation program was drawing fire from critics. For example, Rep. Joseph P. Addabbo, Democrat of New York, said his understanding was that only one major supplier had agreed to comply with the voluntary guidelines. A letter from the head of an independent oil company to Addabbo said that BP Oil Corporation's view "was that they considered the voluntary allocation program to be somewhat humorous."[18]

On June 5, because of the growing criticism of the voluntary allocation program, the Senate passed 85 to 10 the "Emergency Petroleum Allocations Act," which mandated allocations. Action by the House was delayed when, on June 10, Oil Policy Committee Chairman Simon indicated the administration would make a decision on a mandatory program within seven days.

From June 11 to 13, the Oil Policy Hearing Committee held hearings on the voluntary allocation program. Independent marketers and independent refiners generally pushed for mandatory allocations, indicating the voluntary program was not working. Even a few majors admitted that guidelines should be mandatory to assure compliance and lessen contract and antitrust legal problems. James J. Kelley of Kerr-McGee Corporation labeled the present program "ineffective." Murphy Oil Company made it plain they were not going along with the program. Phillips Petroleum Company saw a mandatory program as the only recourse. Standard Oil of Indiana doubted that a significant number of industry members would embrace voluntary guidelines.[19] James Emisen of Oskey Gasoline and Oil Company, a Minneapolis independent, said he had not received one pint of petroleum under the voluntary program and that one of his suppliers told him he did not choose to volunteer to give him supplies.[20]

Ken Catmull, general manager of marketing for Autotronics Systems, told the House Interstate and Foreign Commerce Committee that the voluntary system was doomed to failure because major oil had no business motive to sell products to independent competitors. The majors, he said, could sell all they wanted through their own branded and controlled outlets because of the shortage.[21]

Even though the independents and some majors were convinced of the need for a mandatory allocations program, still the administration did not act. On July 26, more than six weeks after Simon had stated to the House Commerce Committee that a decision was forthcoming on mandatory allocations, the administration had done nothing. Congressman Torbert H. Macdonald, chairman of the Commerce Committee, was disturbed by the administration's lengthy delay. Macdonald's dismay increased when the new energy czar, John Love, notified the Commerce Committee he was unable to testify until the day before Congress' August recess. Macdonald interpreted this as further evidence of the administration's indifference to the problem facing the independent marketer and to the concern expressed by Congress.

Senator Jackson attempted to pass in the last few days before Congress' August recess a "mini" mandatory allocation bill. House leaders were in general agreement and the way seemed clear for passage. However, the House rejected the bill when oil senators attached to it an offensive meat price control provision. Thus Congress left for its summer recess with no mandatory allocations bill.

In November, two months after Congress had returned from its summer recess, a mandatory allocation act was finally signed into law. With the maximum allowable delay, the administration issued regulations on January 15, 1974, which were to become effective February 1, 1974. After approximately one year of supply squeeze, the beleagured independents finally got the mandatory allocations they had been asking for, but perhaps too late to help many of the independents. Furthermore, it did little good for the independents to get their supply back if now the situation was structured so that they could not obtain a competitively priced product.

A question that many people have raised is why the administration did not voluntarily adopt mandatory allocation of gasoline according to the authority that it had received from Congress. Furthermore, why did the administration delay as long as possible before prescribing its mandatory allocation plan when forced to do so by mandate from Congress? Was this related to the influence at the White House of the petroleum companies that were on record as having contributed millions of dollars in the re-election of the President? The giant petroleum companies had worked hard to destroy the price marketers and they certainly did not favor a policy that would force them to continue to sell gasoline to their arch-competitors.

A Banner Profit Year

As competition from the independent discount gasoline marketer was being snuffed out in the final quarter of 1972 and early 1973, the profit picture for the major oil companies turned around. A large contributor to their dramatically improved profits was the diverting of gasoline from the independent price marketer and the selling of this gasoline through their controlled, branded outlets at considerably higher prices. The combination of more controlled volume and higher margin was a big reason for the dramatic change in the profit picture of the industry as 1973 opened.

Quarter after quarter the profits improved in 1973, as can be observed from Table 8–11. For the first quarter, profits for the twenty-five largest petroleum companies increased by 27.1 percent. During the first half of the year, profits were up by 38 percent. For the first nine months of the year, profits increased by 44.3 percent, and for the entire year of 1973, the twenty-five largest petroleum companies reported profits up over the prior year by 52.7 percent.

The return on investment in 1973 was the best that the petroleum companies had experienced since the mid-1950's. The petroleum companies were now in a position to internally generate a much higher proportion of funds needed for huge, capital investment programs facing the industry than at earnings levels of the earlier 1970's. While the petroleum industry needed huge quantities of capital for expanded exploration and development effort to find new oil reserves, it is very sad that this could not have been done within the competitive process and that the efficient discounters had to be made the sacrificial lambs.

Summary and Conclusions

Unable to cope with the fierce price competition from the independent, private brand marketer, the large integrated petroleum companies resorted to what has been referred to in this chapter as the *bold step.* This step was to accept the fact that their marketing approach was hopelessly outmoded. Rather than reforming it to make it more competitive, some giant companies took action that thwarted the workings of the marketplace and destroyed the major source of competition confronting their branded marketing systems.

The largest petroleum company could effect a withdrawal of gasoline supply from the independent, discount gasoline marketer in two ways. It could, on the one hand, reduce its crude oil sales to the independent refineries that sold approximately half of their gasoline to the independent, discount gasoline marketers. The second way in which it could accomplish a withdrawal of supply was to stop its inter-refinery sales of gasoline to buffer companies who in turn sold gasoline to the independent, discount gasoline marketers. Both meth-

TABLE 8–11

*Earnings of 25 Largest Petroleum Companies for 1973
in Comparison to 1972*

Company	First Quarter	Six Months	Nine Months	Year
Exxon	+43.1%	+48.4%	+59.4%	59.3%
Texaco	14.8	28.1	34.8	45.4
Mobil	10.1	25.1	38.4	46.8
Gulf	18.7	46.3	60.0	79.1
California Standard	24.2	33.1	39.7	54.2
Subtotal:	26.0	38.2	48.2	55.0
Indiana Standard	21.5	29.0	32.0	36.4
Shell	49.2	52.0	40.8	27.7
Arco (Atlantic Richfield)	52.2	50.8	37.0	40.3
Sun	42.3	41.6	43.1	48.4
Conoco (Continental)	11.5	17.5	24.1	42.6
Tenneco	14.4	—	12.9	13.4
Phillips	22.1	23.5	30.2	55.3
Union	28.1	35.8	44.9	47.8
Amerada Hess	25.1	41.8	88.3	177.3
Cities Service	17.4	22.0	31.7	36.8
Marathon	49.6	57.8	48.7	62.2
Getty[a]	20.6	52.8	59.5	77.4
Ashland	40.7	37.9	29.1[b]	25.7[c]
Sohio (Ohio Standard)	48.3	80.9	54.7	24.1
Pennzoil	23.9	24.3	27.0	37.1
Kerr-McGee	20.2	21.2	19.9	—[d]
Murphy	57.7	193.8	271.3	239.9
Skelly	2.6	12.2	9.5	17.0
Clark	502.0	649.5	349.7	265.6
Tesoro	—	—	—	52.3[c]
American Petrofina (FINA)	105.1	78.5	57.9	104.6
Subtotal	28.9%	37.8%	37.5%	47.9
Total	27.1%	38.0%	44.3%	52.7

a. Includes Getty's share in Mission Corporation and Skelly.
b. Calendar-year basis.
c. Fiscal year and first quarter.
d. Kerr-McGee earned $462.8 million which was a 21.4% increase (*The Oil Daily*, May 14, 1974, p. 4).

Source: Compiled from the *Oil & Gas Journal;* year-end figures from *Oil & Gas Journal,* February 18, 1974, p. 33.

ods were employed to greatly reduce the supplies available to the discounters.

The abrupt withdrawal of gasoline supplies to the independent marketers beginning in late 1972 and early 1973 seriously crippled the major source of price competition in the marketing of gasoline to the public. Several companies were forced out of business and many more suffered "business shock" from which they may not recover. It will be a long time, if ever, before the independent, discount marketers once again become serious, competitive contenders to the major petroleum companies' nonprice method of marketing gasoline.

The giant petroleum companies have reaped the harvest of their actions in thwarting the workings of the marketplace. As they cut back the supply valve to the price marketers with the occurrence of the shortage, they force fed the gasoline through their owned and controlled branded outlets. Under conditions of shortage, they were able to divert gasoline from the most efficient marketers to the least efficient marketers of gasoline with the cost to the public running in the billions of dollars per year. Unbelievably, the federal government —particularly the executive branch—stood idly on the sidelines and allowed competition to be snuffed out.

The contraction of gasoline supplies from the discount marketers stopped the price warring and price competition which was acting as a severe depressant on the major oil company earnings. As the price discounters lost their supplies, they had no alternative but to increase their prices and to give up their low cost, high volume, low price method of selling gasoline. With the neutralization and destruction of the discounter's competition, the major oil companies were able to curtail price subsidies which had cost them billions of dollars per year and had resulted in corresponding savings to the public. The scuttling of price competition contributed importantly to the 50 percent increase in profits of the twenty largest oil companies in 1973 and to the highest rate of return on investment they experienced since the mid-1950's.

9

Summary, Conclusion, and Recommendations

This study has examined the factors leading up to the petroleum product shortages, how the shortages actually developed and the competitive consequences of the shortages. The findings of the study confirm the public suspicion that certain large companies played a significant role in creating the shortages. They were aided and abetted in so doing by inept management of the oil import program by the executive branch of the federal government.

The petroleum product shortages have resulted in the most rapid rise in the price of petroleum products and improvement in earnings occurring over the past twenty-five years. In the wake of galloping increases in the earnings of the major oil companies, the highly competitive independent marketing segment of the gasoline industry has been severely weakened and its future is in jeopardy.

In several crucial respects, the industry was found to be deficient. First, the industry has failed to build adequate refining capacity in the face of a predictable and balanced growth in demand. Current oil company propaganda campaigns confirm that they had predicted the impending shortage and "warned" the public. These warnings or "threats" were intended to force changes in public policies beneficial to the industry.

Second, the major brand sector of the industry has consistently resisted the motorist's increasing preference for a more diversified, efficient, and streamlined approach to gasoline retailing. They have effectively insulated the retail structure from the market's insistent signals for change. They have persisted beyond the point of reason in protecting their old, outdated, retailing approach from the sovereignty of the driving public.

Third, the large petroleum companies have carefully regulated, and in some cases even sought to destroy, their nonintegrated competitors—especially the more innovative competitors who were threatening the old retailing order.

Competition and the Consumer

There are two basic types of competition in the gasoline industry—*inter*type and *intra*type. Intratype is the competition between similar types of operations. Examples of *intra*type competition would be the rivalry between the major brands of the large petroleum companies such as Shell, Exxon, Texaco, Mobil, etc., and similarly the competition between the private brands of independents, including companies such as Hudson, Fill'em Fast, Martin, Sigmor, etc. Intratype competition among the major brands has focused on costly nonprice marketing tools, including promoting a recognized brand, blanketing markets with large numbers of stations, construction of elaborate facilities for selling a commodity type product, and the use of credit cards, trading stamps, and giveaways. In contrast, intratype competition among the private brands, dominated by the independents, plays down costly frills and emphasizes low cost, streamlined operations where gasoline can be sold to the public at relatively low prices.

*Inter*type competition, in contrast, is the competition between the costly major brand, nonprice marketing approach and the discounting methods of the independent private brands. Intertype competition is concerned with the intense rivalry *between* the dozen or so major brands such as Texaco, Shell, etc., and the hundreds of discount brands of the independents.

Intertype competition is extremely important to the public in two respects. First, it gives gasoline customers the option of either purchasing higher priced, major brand gasoline with all the extras or private brand gasoline at lower prices with fewer extras than the majors. The important issue is that gasoline buyers are given relevant alternatives, and not simply forced to select gasoline from among the closely similar offerings of the majors. Those who want to pay the price for major brand gasoline are entitled to do so, and similarly, those that prefer to purchase private brand gasoline on a lower price discount basis should be given that opportunity.

Second, intertype competition between the major brands and discount brands keeps a ceiling on what Shell or Texaco or other majors can get for their gasoline. It prevents the majors from passing on the full cost of their inefficient marketing style. As intertype competition was destroyed in late 1972 and early 1973, price wars stopped and the major brand retail margins began to rise to unprecedented heights. This loss of margin control is one of the major casualties of the gasoline shortage. The public definitely has a stake in the survival—or, if you will, revival—of intertype competition in this industry.

Major Petroleum Companies in a Bind

The major petroleum companies experienced a deterioration of their highly profitable crude oil operations, both domestically and internationally, and this destroyed their "profit game plan." From 1950 to 1970 crude oil production had been the "profit nerve center" of the petroleum industry. Downstream activities were operated so as to maximize profits at the tax-sheltered crude oil level. It didn't matter at that time if refining and marketing earned only marginal returns or merely broke even so long as petroleum companies could cash in on their highly profitable crude oil production.

By 1971 the era of oil colonialism was coming to an end and the international oil companies were soon to lose their control over world oil prices and the ownership of that oil. The Tehran and Tripoli accords marked the beginning of an era of rapid increases in the price of international crude oil. Following closely on the heels of major price increases were demands from the primary oil producing and exporting countries (OPEC) for ownership of the oil companies' operations within their national boundaries. Over a three-year period of time, some of the countries nationalized the oil companies' holdings and at least took over majority control of the petroleum companies' operations in their countries.

Eastern hemisphere earnings for the major U.S. companies continued a long-term decline as production payments to the host governments rose dramatically. Earnings per barrel dropped from 53 cents in 1962 to 28 cents in 1972. Net Eastern Hemisphere earnings had only risen from $1.2 billion in 1962 to $1.9 billion in 1972, while the gross crude oil production had increased by more than three-fold.

At home, crude oil prices had in essence been frozen from 1957 through the early part of 1973. With increasing cost of production and exploration, domestic production of crude oil was becoming less profitable. In addition the price of natural gas, the companion product to crude oil, was controlled and held at unrealistically low levels given the prices of other fuels.

The need for downstream profits by the integrated petroleum companies came at the wrong time. It occurred just when the modern, sophisticated, self-service operations were being pioneered by the independents. These new intertype competitors struck at the very heart of the nonprice major brand marketing approach. The self-servers had a new set of economics with lower costs and lower prices while the majors' style was increasing in cost. The entire structure of 200,000 plus conventional, major brand retail service stations was in jeopardy. Repeated attempts by the majors to increase refining and marketing margins in 1970 to 1972 (until August 15) resulted only in price wars and increased market shares for the price marketers. Steps by the majors to improve their downstream profits came too late and were inadequate to the task. Secondary branding, market withdrawals, rack pricing, and self-service experiments did not improve the plight of the major oil companies. Downstream profits remained inadequate and the majors' market share declined. The majors either had to change their marketing style or destroy their intertype competitors. They chose the latter course.

Engineering and Application of the Shortage

In early 1972 the majors began to build toward a moderate product shortage by underutilization of their refining capacity and thus drew down refined product inventories. The record shows a distinct break with past seasonal operating patterns. No plausible alternative explanation of this abrupt break with the past has ever been given.

If refineries had been operated at higher capacity during 1972 and the first five months of 1973, refining capability would have been adequate to meet the demand for home heating oil and gasoline through the summer of 1973. The heating oil shortage of the winter of 1972–73 was the direct result of low refinery capacity utilization during 1972. Refinery output east of the Rocky Mountains for the first four months of 1972 was actually below that for the same period during 1971. Examination of individual refinery runs of companies showed that several of the larger U.S. refineries were primarily responsible for the industry's low output during the first part of 1972. The underutilization of refining capacity during the first five months of 1973 aggravated the already tight petroleum products situation that had been created and gave rise to the gasoline shortage during the summer of 1973. At this time the output of several independent refineries was down, since the large integrated petroleum companies had reduced crude oil sales to them.

The underutilization of refining capacity in late 1972 and in early 1973 was adversely affected by the federal government's handling of the crude oil supply situation. During 1972—especially during the second half—several U.S. refineries were desperately short of needed crude oil. Domestic production peaked in early 1972 making it necessary to rely more and more on imports. The regulation of the level of imports was the responsibility of the executive branch of the federal government and this task was badly handled.

The federal government failed to see its need to assume a "balancing wheel" role that had previously been performed by Texas and Louisiana in their prorationing activities. Once these states went to maximum allowable production, the federal government needed to increase the level of imports to make sure all domestic refiners would have adequate feedstocks. The fact is that the mismanagement of the Oil Import Program during 1972 resulted in a virtual "starving" of the lesser majors and the independents of needed supplies. Inadequate and delayed increases in the level of imports were made three times during 1972. In every case, they underestimated the needs of domestic refiners.

In partial defense of the responsible government policy-makers in dealing with the crude shortage, it should be noted that they were receiving mixed advice from different segments of the industry. The independent refiners and the crude-starved majors were pleading for increases while the numerically and politically powerful independent crude producers and the crude-rich majors counseled restraint in increasing imports.

In addition, the aimlessness of oil import policy in the early 1970's created considerable uncertainty for management decision making and curbed needed expansion in refining capacity. Only after the Oil Policy Committee was transferred to Treasury Department leadership and oil import policy was clarified to emphasize crude over finished product imports did any significant new refinery construction begin.

While the initial shortage was created by a programmed underutilization of refinery capacity in order to pressure certain changes in government policy and competitive conditions, by late 1973, the shortage was due to inadequate refining capacity. That is, it was a manufacturing shortage rather than a manufactured shortage. It was obviously deepened and made more severe by the Arab crude embargo. Nevertheless, it would have existed without the embargo.

Focusing Shortage on Independents

The mechanism by which supply was denied the independents is simple. Once there was a shortage—intentional or otherwise, the large petroleum companies merely cut off inter-refinery sales of gasoline to the buffer companies (smaller majors, independent refineries, and superbrokers) who traditionally sold gasoline to the price marketers. This was accompanied by a reduction in crude oil sales to the independent refiners who also were a prime source of gasoline for the discounters. By mid-1973, many independent refineries were operating at less than 70 percent capacity because of insufficient feedstocks. Yet the majors were running at 95 to 105 percent of capacity.

This strategy of focusing the shortage worked. By October 1973, the major brand market share was up nearly 7 percent and at much higher prices, while the nonmajors were down by nearly 4 percent. Price wars also came to a halt. Prices which had gyrated wildly from normal levels during 1971 and 1972 stabilized abruptly on August 15, 1972. Without adequate gasoline, the discount competitors no longer had either incentive or capacity to challenge major company price restorations. Ironically, this date, which marks the end of intertype competition in the marketing of gasoline, was the second anniversary of President Nixon's price control program. Since that time retail margins have risen to unprecedented heights, even under stringent price control.

Profits dramatically increased as intertype competition, and in turn, price competition, was neutralized and destroyed. After all, this was the purpose of destroying intertype competition in the first place. Nineteen hundred seventy-three was the banner year for "big oil." In fact, it should have been embarrassing that the oil industry profit so dramatically from a shortage that they had helped engineer and that they had carefully focused on their intertype competitors. The driving public will pay billions of dollars per year for this shortage, and it is all justified in the name of free enterprise and the profit system. But whatever happened to the public's right to competition?

Conclusion

It is one thing to make the case for change. It is quite another to responsibly detail what changes should be made. The specification of a complaint does not necessarily imply its solution. There is no question but that this is a complex industry, and the solution to a problem in one area may create even greater problems in another. This "system effect" is especially troublesome in an industry which is as basic to our economic and social life as the energy industry.

But while this is true, the very importance of the industry means that we dare no longer leave such key decisions to the "enlightened" self-interest of a few dozen self-appointed executives who seek to define the public interest from the remote vantage point of their executive suites. Such decisions should be reserved for the people or their duly elected representatives. They should only be made by private executives when they are subject to the detailed control of a workably competitive market.

"Corporate responsibility" has been put to the test in this industry and has been found wanting. The giants of the industry collectively dictate how and at what price gasoline is to be marketed, instead of responding to varying consumer desires for different combinations of products, services, and prices. What is good for Standard Oil does not seem to be what is good for the country.

What is wrong with this industry? Why has it not responded to customer desires for different combinations of product, service, and price? Why the almost total reliance on a high price strategy built on product differentiation, intensive market coverage, and huge sales promotion expenditures? Independent private brand marketers of gasoline sell much higher volume per station—often two to three times as much as the major average—and at considerably lower cost and prices. With only a slight reduction in convenience, no loss of time in being serviced, and improvement in quality of service, gasoline could easily be marketed through one-third to one-half the present 220,000 service stations,* at much lower cost and price. In a competitive industry, this adjustment would have long since been made. Obviously, the answer is that the industry is not competitive. How else would such a wasteful and unresponsive arrangement be sustained?

Traditional Indices of Competition

The traditional indices of market power applied to the petroleum industry would not signal the absence of competitive behavior in this industry. As pointed out in a 1973 Trade Commission report, it is not an excessively concentrated industry. On a national basis, no marketer has as much as 10 percent of the market. Even in specific metropolitan markets, no single seller has a market share anywhere approaching those of other industries which seem to be workably competitive. In most markets, the business is shared by ten to fifteen important retail sellers; the cost of retail entry is not high and product differentiation is weak.

The same is true at the refining level. There are 150 refining companies in the United States. No important geographic area is served by less than ten refineries. There are no special trade secrets or patent advantages held by any small group. The technology is well understood and available to all. In addition, most of the refining capacity is actually constructed by large independent contracting and engineering firms that will build for all comers.

Nor are economics of scale or capital costs a real limit to entry. One hundred fifty thousand (150,000) barrels per day refining is

---

* The Survey of Current Business defines a service station as an outlet doing 50 percent or more of its business from the sale and service of petroleum products. This definition considerably understates the number of branded outlets, for the ten largest companies alone have more than 220,000 outlets.

generally referred to as the efficient size. Such refineries can be built for about $250 million. This is a sizeable but not a major capital cost, given the American capital market. Besides, many, if not most, refineries now operating have capacities of only a fraction of this amount.

Even the market structure at the crude production level does not appear to be overly concentrated if it is looked at in isolation. There are over 7,000 producers of crude. The big eight in the major oil-producing Gulf Coast states produce only 52 percent of domestic crude oil.[1] Exxon, the largest domestic producer, controls only 10 percent.[2] No one has a monopoly on the methods of finding new oil. The techniques are known and available to all. Admittedly the big firms have an advantage because of the huge capital requirements, but their advantage is not absolute.

The structure of the international crude market is substantially more concentrated than the domestic market, but looked at in isolation, not excessively so. True, it is largely controlled by the so-called seven sisters, but others are also active in this area and their share has grown substantially since 1960. Again, no single firm controls more than 17 percent—the big four control 56 percent and the next four only 25 percent.[3]

Transportation is the sector of the industry that is least competitive. Pipelines in particular are a problem. Crude oil gathering lines from a specific oil field are often owned or otherwise controlled by one to three buying companies. Major intrastate and interstate crude and products trunklines connecting producing and refining and/or marketing areas are primarily joint ventures of small groups of companies, or are owned outright by single petroleum companies. This follows from the fact that pipelines are subject to important economics of scale. A single pipeline of a given capacity is substantially cheaper to build and operate than several small parallel pipelines. However, even here there are some alternate modes of transport, such as barge, tanker, and truck, that are not so controlled. If the demand were great enough and the monopoly profits attractive enough, even pipeline entry is not out of the question. In fact, a few independently owned (true common carriers) pipelines do exist and prosper.

The structure of the industry at each level, thus, does not seem to be overwhelmingly anticompetitive when each is considered in isolation. Therefore, the documented anti-competitive behavior must be due to other reasons. It is the thesis of this study that the *poor performance* record is the direct consequence of the way the industry is organized, i.e., *it is largely due to the vertically integrated structure of the petroleum industry.* If each level of the industry were free to act in the context of the market structure at that level, its behavior would be substantially different from what it is when forced to act in ways that maximize the profits of the entire vertically integrated system. How can this be?

Vertical Integration—Economic and Legal Views

The anticompetitive role of vertical integration is somewhat hard to grasp. Economic theory suggests that if each level of the vertically integrated industry is workably competitive, then all levels should also behave competitively.[4] True, if one level is not competitive, then the market power at that level can be extended to otherwise competitively structured levels by means of vertical integrations. But even in such cases the problem is one of market power at the particular, offending level, rather than of vertical integration.

Most economists regard the anticompetitive complaints against vertical integration as groundless.* They hold—and properly so—that a vertically integrated firm cannot earn double profits as is often charged. Neither can it gain unfair advantage by selling raw material inputs to itself from its affiliated companies at cost or by marketing its finished products through its own retail outlets at less than cost. Economists claim the subsidization argument does not hold up. Many economists would even argue that vertical integration confers no ability to alter market price or impede entry. To this group, practices conventionally labeled exclusionary appear to be either tactics equally available to all firms, or only a way of maximizing returns from a horizontal market position already held.

The dominant view of economists seems to be that the vertically integrated firm is no different from other firms in its ability to compete. It cannot gain unfair advantage by obtaining goods at cost from

* Exceptions to this belief would be the case of the monopolist who cannot discriminate in price effectively without control of his outlets or the case where there is clear intent to destroy competition accompanied by behavior that confirms that intent.

earlier stages or by foregoing profits on a part of its operation or by selling products below cost. The manipulation of profits between production and marketing stages is held to be merely an accounting illusion. In this respect, the vertically integrated firm is no different from any growing multiproduct, multimarket firm. The established always subsidizes the new. The multiproduct or multimarket firm—and there are very few single product, single market firms—has advantages similar to those which allegedly follow from vertical integration. Thus these economists hold much of the behavior associated with vertical integration—charged by some to be predatory or exclusionary—to be the essence of competition.

The view of the current Justice Department staff, and the view generally held by the courts, conforms to this conventional economic wisdom. From the beginning, the courts have held vertical integration by ownership (as opposed to contract) to be legal unless it is used as an instrument of unfair competition.

In the 1920 U.S. Steel case, the Supreme Court said, "A vertical combination actuated by considerations of efficiency and marketing and not by a desire to create a monopoly is not in violation of the Sherman Act."[5] Until 1940 the courts rarely attacked vertical integration as such. In many of the landmark cases before 1940, vertical integration was present but left undisturbed. For example, in the Standard Oil, Harvester, and Corn Products cases, vertical integration—though important—was not condemned. Nor was vertical disintegration part of the remedy sought.

After 1940, however, vertical integration was attacked in Pullman cars, petroleum pipelines, aluminum, food chains, taxi cabs, movies, and steel. Quasi-vertical integration in the form of requirement contracts came under increased attack and, with the 1950 amendment of the Clayton Act, vertical mergers were increasingly frequently contested. But, even so, sharp disputes remain over both the effects and legality of vertical integration. There has clearly been no judicial groundswell for divorcement.

There is good reason for the courts' continued resort to a "rule of reason" as opposed to a *per se* approach to the regulation of vertical integration. Consider the case of a manufacturer integrating forward into wholesaling and retailing. Such a move could be viewed as a method of foreclosing competitors from an opportunity to sell in particular markets. Or, it may be merely an effort on the part of the

manufacturer to eliminate wasteful and inefficient intermediate transactions and to gain more effective retail promotion and distribution.
Which is it? The question cannot be answered in the abstract. The
courts properly have sought to distinguish between cases where vertical integration seeks to foreclose markets and destroy competition
and cases where vertical integration merely reflects healthy expansion or an attempt to introduce production and distribution efficiencies. This is why the law is so cloudy here.

Ultimately, the legality of vertical integration must depend on
the effect or purpose of the arrangement, on the one hand, balanced
against various strategies and competitive factors in the markets
affected by it on the other. The superior ability of the integrated firm
to compete clearly poses a threat of unfair competition for the nonintegrated firm; it also creates an opportunity for lower end-prices
to consumers. This is the dilemma.

Vertical Integration in the Petroleum Industry

Against this background, how can vertical integration be held
to be anticompetitive in the petroleum industry? The answer lies in
the "conduct" and "performance" of the industry. Conduct is concerned with whether the power inherent in vertical integration has
been used for anticompetitive purposes. Performance focuses on the
consequence of the conduct and whether or not it serves the public
interest.

Since World War II the "profit nerve center" of the petroleum
industry has been in the crude oil divisions of the integrated petroleum companies. As a consequence, relatively little profit has been
made by the marketing, storage, or refining activities of the large,
vertically integrated petroleum companies. The impact of this "conduct" has been to place downstream, nonintegrated competitors—
refineries, terminal operators, and marketers—at severe competitive
disadvantage.

Since 1950 the integrated oil companies have taken over several
of the important independent refineries and there have been built
no new independent refineries with over 50,000 barrel per day capacity. With artificially high crude oil prices and low subsidized wholesale product prices, the squeeze on the independent refineries has
been very effective.

Most of the independent terminal operators were driven out of business in a similar manner and many of their operations were taken over by the large integrated companies. For example, during the 1950's and 1960's the number of independent terminal operators in the Midwest was estimated to be reduced from more than 80 to less than a dozen.

Also during the late 1950's and early 1960's, a majority of the largest chains of discount marketers were absorbed by the integrated oil companies. Included were such independent brands as Douglas, Wilshire, Hancock, Regal, Harbor, Oklahoma, Perfect Power, Gasateria, Bulk, Superior, Western Oil and Gas, Kayo, Cosden, Spur, and Direct.

Only after 1965 did the independent discount marketers stage a major comeback. The reasons for the reversal include the general price restoration of 1965 and the moratorium on widespread price wars for approximately five years. In addition, in the latter 1960's and early 1970's the independents adopted the lower cost self-service method of marketing at the same time the majors' method of marketing was becoming more costly. This increased the competitiveness of the independents.

The question now to be examined is whether or not this conduct resulted in a "public gain or loss." If the conduct resulted in lower prices and increased consumer choice, then the argument could be made that nothing should be done to restructure the industry. If the reverse is true, the petroleum industry should be restructured to increase competition in order to improve its performance.

The fact is, the driving public has not benefitted from the conduct of the petroleum industry. As noted in Chapter 4 and also in *Competition Ltd.*, the integrated petroleum companies' method of selling gasoline has been extravagant and inefficient. Gasoline marketing policies and practices have led to massive overinvestment and overbuilding of branded service stations. The market for gasoline has been so fragmented that it has greatly increased the retail cost, and hence the price, of gasoline sold to the public.

Gasoline could easily be sold through one-half of the present 250,000 gasoline stations as noted by the chairman of Gulf Oil before security analysts in 1974. With greatly reduced investment and operating cost, the price at which gasoline is sold could be significantly reduced. Until the shortage, many independent, private brand marketers sold a much higher volume per station—often two to three times the major brand average—and at considerably lower cost and price per gallon. It is estimated that, if the number of major brand service stations were approximately halved, prices could be reduced from 3 to 5 cents per gallon. The savings to the public would be from 2 to 3.5 billion dollars per year. In a competitive industry this adjustment would long ago have taken place.

In addition to overinvestment in branded outlets, the structure of the petroleum industry has also resulted in underinvestment in refining capacity. Even without the Arab oil embargo, the United States would still have experienced shortages of petroleum products because of inadequacy of refining capacity. The structure of the industry effectively prevented independent refineries from generating the capital necessary to expand refining capacity and blocked new entry. Meanwhile, the integrated oil companies were unwilling to make the needed investment in expanding refining capacity to keep up with demand in the United States.

Proposals for Changing the Petroleum Industry

Calls for reforming the petroleum industry have become much more numerous with the escalation of product prices and the marked rise in profits while the public has suffered the consequence of the shortages. It seems very likely that something will be done in the near future, and it is important that the action taken benefit the long-run public interest. It will do no good to be stampeded into accepting a proposition for changing the petroleum industry which is either so watered-down it will have no effect or so radical it will destroy what is good as well as what is bad in the petroleum industry.

There are three major types of proposal being advanced for reforming the petroleum industry. One type of reform would have the federal government operate or regulate the petroleum industry. A second type would place the government in the industry as a direct competitor. The last type of reform is concerned with the restoration of the competitive process as the regulator of the industry.

Government Operation and Regulation of the Industry

At one extreme are proposals suggesting that the only way to correct the dysfunctioning of the petroleum industry is nationalization. Those proposing a nationalized petroleum industry are assuming that the government would be more responsive to the public interest and would be able to allocate resources more efficiently in the industry. However, there appears to be little ground for supporting such a conclusion. Nationalization usually has been attempted in quasi-capitalistic societies, in relatively simple industries where business practices are relatively straightforward. Business arrangements in the petroleum industry are highly complex. Conditions and circumstances involving the operation of the industry are continually changing. The obstacles to the successful nationalized operation of the petroleum industry are enormous. Thus nationalization should be viewed as a last resort.

There are many Congressmen who feel that the best way to deal with the problem of the petroleum industry would be to subject it to some form of public utility regulation. One measure involves the federal chartering of the petroleum companies. This would require the petroleum companies to disclose more of their internal corporate decision making. Also a public representative would participate in all major policy decisions. Such a proposal would only mildly constrain the freedom of decision making of the petroleum companies. It would be primarily symbolic and would change very little.

Other proposals for public utility regulation involve direct public control over price, product mix, and expansion of productive capacity. Under this form of regulation, prices and output would actually be established by public authority after formal hearings before a public commission. There are many problems with the public utility approach. The U.S. experience with regulated industries has not been exceptionally good—for example, the public utility regulation of the transportation industry has certainly not been a happy experience. It is very unlikely, given the complexity of the petroleum industry, that a regulatory agency would be able to improve upon its performance.

Government as Competitor

Several proposals would have the federal government enter the industry as a competitor. Some of these proposals recognize the barriers to entry in the refining of petroleum products. Others stress the importance of government intervention in the production of crude oil.

Senator Phillip A. Hart, chairman of the Antitrust and Monopoly Subcommittee of the Senate, has introduced legislation requiring the federal government to build seven large, independent refineries throughout the United States. These refineries would in turn sell badly needed gasoline to the independent, discount gasoline marketers and fuel oil to independent distributors. Eventually these government financed refineries would be sold to private corporations.

Hart's proposal overcomes the entry problem in the refining business, but it does not deal with the basic problem of the independent refiner. Over the years the independent refinery segment of the industry has been squeezed between low product prices and artificially high crude oil prices. There simply was not enough margin at the subsidized refining level of the integrated petroleum companies to make independent refinery investment pay. It makes little sense for the federal government to subsidize the construction of independent refineries if they are going to face the same squeeze that weakened independent refineries in the past.

Senator Hart and others have also proposed a federal oil and gas corporation that would actually explore for oil and gas on government lands. This corporation would produce and sell crude oil to the seven government-built refineries. Any surplus crude or product could be sold to other refiners.

The exploration and production of oil and gas is a highly sophisticated business undertaking. One would have to question whether the government could assemble the skilled individuals and organization structure needed for this task, or whether they could do the job as efficiently as the private oil companies. Actually, the shortage of crude oil in the United States is largely the result of government policy rather than any unwillingness on the part of the oil industry to explore for oil and gas. Less than 5 percent of the outer continental shelf has been leased to date. This is the area where most of the better prospects for new oil discoveries exist. New policies for more rapid leasing of the outer continental shelf rather than government entry into the industry would speed the development of oil in the United States. In addition, a different method of leasing federal oil lands must be developed. The present bonus bidding system favors the large companies, is anticompetitive, and drives up the cost of oil production.

Restructure Petroleum Industry

A third major category of proposals for reforming the petroleum industry are those that would restructure the industry. Such propositions seek to maintain the private enterprise system, but to remedy its defects. The general idea is that limited government intervention can restore the competitive process as the principal regulator of the petroleum industry. In 1911 the Standard Oil Trust, with its monopoly in refining that extended backward into pipelines, was broken into 33 companies in order to restore competition in the industry.[6] Sixty years later strong sentiment indicates that major surgery is once again needed in the petroleum industry. Presently the concern is with the market power of the huge petroleum companies resulting from the high degree of vertical integration that has characterized the industry.

Several divestiture proposals have been made recently. One proposal would separate refining and marketing. The question that must be asked about such propositions is whether they attack the symptoms or the cause of market power. Consider the following scenario:

1. No refiner is permitted to own or otherwise control retail gasoline operations. Present operations must be spun off.
2. Competition becomes much more intense at retail since entry is easy and product is available to all on equal terms. Chain retailers will bid the supply price of gasoline down as they play off one refiner against the other. Retailing will become more responsive and efficient.
3. With an effective lid now placed on refiners' prices by competitive retailing, the integrated producer-refiners are inclined to raise crude prices to maintain their programmed profit growth.
4. The nonintegrated refiner now squeezed by rising crude prices and stable or declining product prices faces declining margins and is inclined to withdraw from the market. Prospective nonintegrated refiners are reluctant to enter the market.
5. Refining becomes more concentrated. Retail countervailing power relative to refining is thereby reduced and wholesale product prices rise to restore refining margins on top of the rising crude prices.

6. Retailing is now operating efficiently at low margins, but retail prices continue to rise as integrated producer-refiners continue to raise their prices of crude and refined product.

On the other hand, depending upon how the major brand problem is handled, there might not even be much improvement in marketing efficiency. If refiners were allowed to continue to maintain their brand identity and to franchise outlets to sell their brand, it is unlikely that many stations would sell more than one brand. If this were so, a less than arm's-length relationship would continue between branded retailers and their suppliers. Under such circumstances gasoline marketing might look and behave much like the present arrangement, even with divorcement.

The gains from the separation of refining and marketing do not make this an attractive approach for improving the competitiveness of the petroleum industry. Retailing might or might not become more efficient and responsive, but in either case the rest of the industry would continue to exploit its economic power. The nonintegrated independent refineries caught in the middle without a strong crude oil base would probably pass into oblivion. A primary source of supply of gasoline for the efficient independent discount gasoline marketer would be eliminated.

Marketing divorcement would not resolve the key problems of the industry. In fact, if marketing were divorced without other reforms, it could further increase the market power at the production and refining levels of the large integrated petroleum companies.

Other proposals call for breaking up the petroleum industry at several levels. It has been proposed that in addition to divestiture of marketing that separation also occur between refining and pipelines and between pipelines and crude oil operations. The more ambitious programs divide the petroleum industry into four independent segments—marketing, refining, pipelines and transportation, and crude oil.

Multiple level restructuring of the petroleum industry needs to be carefully thought through. Effective and efficient organization and co-ordination of the petroleum industry could be destroyed by over zealous divestiture schemes. As previously argued, a vertically structured industry with integration of certain activities can achieve greater efficiency and public benefit than an industry void of integration. In fact, rivalry between integrated and nonintegrated segments of an industry becomes so powerful that it overwhelms and destroys other competitive segments, and that power is buffered and protected by layers of vertical integration.

Recommendation

Central to the recommendation that follows is the belief that a free and open market subjected to the inexorable, regulatory forces of competition is the most effective and most efficient means of causing this vital industry to behave in the public interest. Structure governs behavior. To the extent that the behavior of this industry is suboptimal from a standpoint of public interest, it is because the specific market structure within which the decision makers function fails to channel their self-interest in the desired direction. It is a failure of structure, not of motives. Consequently, our primary recommendation is a reform of structure, which seeks to restore competition as the basic regulator of the key economic decisions of the industry. The market has no mercy; regulators do. Our recent experience shows regulators permitted prices to increase more in six months than the market had permitted in twenty years. They did this in the name of price stability and in the furtherance of the consumer's interest.

It has been argued throughout this book that the major competitive problems of the petroleum industry lie in its vertically integrated structure which has led to its wasteful and unresponsive performance. Excess investment in retail facilities, emphasis on extravagant nonprice means of competition, reluctance to expand refining capacity in the face of a balanced and predictable growth in demand are all due to vertical integration.

If one accepts the anticompetitive nature of vertical integration in the petroleum industry, then the most obvious reform would be to eliminate it by requiring vertical divorcement of presently integrated operations. Actually it is that tie between crude production and refining that is crucial. Without crude power, the majors would not have refining power. And without refining power, they would not have marketing power. Destroy or isolate the power of crude and the whole house of cards will fall. Independent refiners would enter the industry and existing refineries would expand if they could procure dependable crude supplies on equal terms with their other competitors. If the refined product from independent refiners were available, independent marketers would stage a comeback and vigorously challenge the costly marketing approach of the majors.

Even integration between refining and marketing by the present majors would pose no serious problem for competition. As noted, none has over 10 percent of the market. The integrated major refiner-marketer would have no advantage not possessed by the independent refiner-independent marketer channel. He would have to pay as much for his crude as the independent refiner and he would have no special claim on crude supply in time of shortage. Competition would no longer be unfair and manipulated squeezes by integrated competitors would rapidly diminish.

One additional gain would follow from the separation of crude production from refining and marketing. While this move would not increase the degree of competition between crude producers, it would create a countervailing power in the form of independent buyers who would seek to bid prices down and otherwise resist crude price increases.

This separation between crude and refining must be achieved by legislation. It is a political rather than a judicial matter. Present antitrust policies are not able to cope effectively with market power as it exists in this industry. Between 1890 and 1955, the courts have only ordered dissolution, divestitures or divorcement in twenty-four litigated Sherman Act cases.[7] And between 1955 and 1968, only two more.[8] Recent cases typically involve vertical integration through merger.

This desired divestiture of crude production operations can be achieved in two ways. It can be ordered or it can be induced. If it is ordered, it can be handled in ways similar to the dissolution of the public utility holding companies under the Public Utility Holding Company Act of 1935. This was one of the most successful pieces of industrial surgery in the modern era. Essential equities were preserved through due process, but the public interest was speedily and effectively served. In that case a special division was set up in the SEC. There is no reason why a Petroleum Divestiture Act could not be so administered. Court-appointed referees, much like those common to bankruptcy proceedings, could be used to oversee the protection of the equities connected with the divestiture. In cases of ordered divestitures, the administrative rulings would be subject to traditional judicial review.

It would also be possible to induce divestiture by denying integrated crude producers the tax advantages now enjoyed by all crude producers. Such legislation, however, might be subject to attack on constitutional grounds. In any event, it assumes a continuation of present tax preferences which in our opinion is undesirable as class legislation.

It is the recommendation of this study that: *The vertically integrated structure of the industry should be modified to prevent common ownership or control of crude and refining.* This is the tie that binds. With freely available crude in an open market, decisions by existing refiners to restrict expansion would be countered by new entrants in refining who believed that money could be made from a growing product demand. And with adequate refining capacity, free entry into marketing is assured.

Final Statement

The questions about how much, to whom, and at what price petroleum products are to be made available to the American people is a matter of major concern. Basically three options are open to us.

One option would be to let the decisions be made by private managerial decree, that is by the major oil company executives who will decide what is best for us. No doubt, these executives are all good-hearted men, but this option no longer has much appeal, given the sorry record documented in this book.

A second option would be for the decisions to be made by public servants in the governmental bureaucracy. The record of accomplishment in the regulation of economic activity by public officials is not any better than that in the first option. In fact, if history is a guide, it will not take long for the regulators to team up with the industry and regulate it in their interest instead of ours. Look only at transport, agriculture, banking, housing, securities, and others to see how this "marasmus" unfolds.

The third option is the one this book supports. It is the option of letting the people decide for themselves through freely competitive market processes. Competition can be made to work as an effective and sensitive regulator in this industry. We would all be much better off if it were allowed to function.

The Growth of Self-Service

The price and general competitiveness of the independent marketers were greatly enhanced through their pioneering and developing of self-service gasoline marketing. The independents had always stressed customer savings and quick service. They had operated on a discount formula of selling private brand gasoline through relatively few centrally located stations on a high volume, low unit cost, and low price basis. As they spearheaded the development of lower cost self-service gasoline marketing, the independents were often able to increase their discounts by approximately 2 cents per gallon, which frequently meant savings to customers of 5 to 7 cents per gallon in comparison to major brand service station prices.

Before the advent of self-service, the major oil companies, through a sophisticated method of price subsidizing (called price protection), had been able to focus their limited price response in the vicinity of strong independent marketers and had been able to hold the independents' growth within tolerable ranges. However, the majors were unable to maintain a check on the growth of the private brand marketers as the independents rapidly adopted self-service with its new set of economics. As self-service was added to traditional independent discount marketing in the late 1960's and early 1970's, the independents grew rapidly at the expense of the costly, nonprice major brand method of marketing. Thus, to a considerable extent, this more efficient and, to some, more attractive method of marketing broke the back of the expensive, inefficient, archaic method of major brand marketing.

The independents' development of self-service would have been problem enough if it had come simply from traditional discount operators. However, the flames of self-service were spreading through a new breed of independent price competitor in addition to those already existing. Several of the new breed looked upon gasoline simply as another product line, and this made them even more of a competitive threat.

Growth of Self-Service

Today's self-service had very humble beginnings that can be traced back to the latter 1950's and early 1960's. By the mid-1960's only a smattering of self-service existed in markets in the near western part of the country. But, like a slowly developing storm, self-service gasoline rolled into a large number of markets with great impact in the latter part of the 1960's and early 1970's, and the market share of the independents rapidly climbed at the expense of major brand marketing.

Some idea of the market potential for self-service can be obtained from considering its rapid penetration into some of its first markets. By 1971 various estimates placed self service at 40 percent in Boise, 30 percent in Phoenix, and 35 percent in Salt Lake City.[1] Mr. J. Whiting, an independent gasoline marketer from Phoenix, stated in 1972 that "close to 60 percent of the Phoenix market is now self service." This approached the prediction of a spokesman for American Oil Company of a "possible . . . market share in the high 50's."[2] In larger market areas estimates placed self-service at 20 to 28 percent in Arkansas[3] and 35 percent in east Texas by 1972.[4] In 1972, five years after being legalized, self-service was also estimated to account for 25 percent of Wisconsin gasoline sales.[5]

Self-service had to fight its way beyond legal barriers as it grew. Pat Griffin, a leading pioneer of self-service, found legal obstacles to his Gasamat chain in the Rocky Mountain area. It was said that Griffin would enter a new area with a petition for injunctive relief in his back pocket. After victories in the near West, legal barriers to self-service were knocked down throughout much of the rest of the country by the efforts of other early developers of self-service. In 1968 only 27 states allowed some method of self-service dispensing of gasoline. By 1971 the number had increased to 35 and by 1972 only 8 states prohibited self-service.

In 1969 self-service was estimated to account for only one percent of the gasoline sold in the United States, but by 1972, according to *NPN* statistics, self service accounted for approximately 7 percent of all gasoline sold. As shown in Figure A-1, the shortage is forecast to retard the growth for a period of time but *NPN* predicts that by 1980 self-service will account for approximately 42 percent of the gasoline being dispensed.[6] Perhaps the ultimate market share of self-service in the United States may approach the 60 percent level of Sweden.[7]

Price and Other Appeals

The initial success of self-service was primarily based on the appeal of an extremely low price. For example, Pat Griffin's early Gasamat units were as barebones as one could devise. To hold down the cost of real estate the locations selected were often low-cost secondary locations. Investment in property improvement was extremely limited and was accomplished by such techniques as gravel driveways, concrete block buildings, and hand painted signs. Furthering this low cost orientation was the use of live-in, retired and elderly people who were paid a small commission to sell tokens and to act as watchmen over the operation.

The low cost formula of Gasamat's operation made it possible for

FIGURE A-1

Self-Serve Share of U.S. Retail Gasoline Market (estimated)

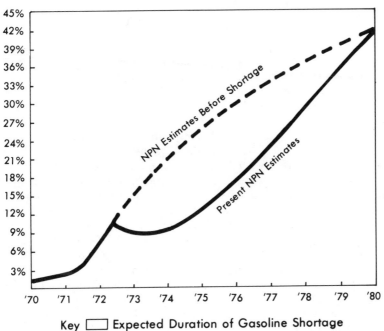

Key ☐ Expected Duration of Gasoline Shortage

Source: National Petroleum News (September, 1973), p. 98.

Griffin to generate a high rate of return on investment with a five-cent operating margin. This five-cent margin would contrast to comparable major brand marketing margins of 10 to 14 cents and permitted Gasamat to sell gasoline often for 5 cents or more under the major brand service station price for gasoline.

Another low cost technique for self-service dispensing of gasoline was to tie in gasoline with an existing business establishment. Since existing properties were more fully utilized, there was no incremental ground cost charged to the gasoline part of the operation. The most frequent type of gasoline tie-in has been with convenience groceries and dairy stores, but other tie-ins exist with drive-in restaurants, liquor stores, and real estate offices. The incremental investment for adding gasoline to existing retail stores was also low, costing about $10,000 to $15,000 for one island having two gasoline pumps, a price sign, and remote control equipment. This low investment, self-service approach would compare to a total investment for land and improvement for new, major brand stations frequently costing $200,000 or more.

With the minimum investment in facilities and little or no incremental cost attributed to land and labor, the tie-in units were extremely efficient. Several of the larger companies with tie-in self-service gasoline had total operating margins of under 5 cents per gallon, less than half the margin of their major brand competition.

The 4- to 5-cent margin of self-service operators would be in contrast to 6 to 8 cents per gallon for traditional independent private brand operators and to 10 to 14 cents per gallon for major brand marketing. The differences in operating margins gave the pioneers of self-service a substantial price discount from the major brand level and a lesser but still significant price advantage relative to the traditional private brand operators.

The self-service tie-in added another dimension to the appeal of self-service—convenience of location. Many of the tie-in units were located in neighborhood areas where drivers made quick purchases of groceries or dairy products, alcoholic beverages, and take-out food. Besides the convenience of location another advantage of many tie-in operations was their long hours. Most of the retail stores to which self-serve gasoline was added operated from seven in the morning until eleven at night, or even 24 hours a day. Self-service picked up some of the off-hour business that many branded service stations were ill-suited to handle.

Had price been the only feature of self-service, the competitive threats which it presented to the major brand method of marketing would have been quite limited. As self service matured, however, it was significantly upgraded and a quality approach developed. Self-service took on an air of respectability with selection of better locations and the building of some very attractive facilities. So, in addition to price and convenience, self-service began to reach the quality-conscious, economy-minded customers, greatly enlarging its marketing penetration potential.

These upgraded facilities also offered customers other benefits. At a spacious, well laid-out self-service unit, drivers could quickly purchase gasoline if they were willing to serve themselves. This was a real benefit for customers who did not want to wait for slow-moving or otherwise occupied attendants. Another feature of self-service for some customers is that they could look after their own cars and service them the way they wanted, rather than having to rely upon poorly paid, indifferent attendants.

Early Self-Service

There was a short-lived surge of self-service almost two decades before the explosive growth of self-service that occurred at the end of the 1960's. Large-scale self-service gasoline operations were first introduced in the Los Angeles Basin on May 1, 1947, by Urich Oil Company.[8] Within a few months of opening, Urich's first self-service outlet was selling nearly one-half million gallons per month—50 to 100 times other stations' volume at that time. Others joined in the self-service development, such as Powerine, and a number of stations pumped over 100,000 gallons per month, which was staggering volume by standards in those days.

Within two years of the opening of Urich's giant self-serve station, self-serve was estimated to control more than 5 percent of the gasoline sold in Los Angeles stations, and there were many more outlets being constructed. Starting in early 1950, self-service found itself embroiled in a severe price war with its major brand competitors. Instead of selling for a nickel per gallon less than the majors, the price advantage was cut to one cent or none at all. Adding to the pricing problem were supply difficulties experienced by Urich and Golden Eagle, and together these factors brought to a close the first glamorous era of self-service.

The growth of self-service was very rapid for a new marketing concept, but the retreat was even quicker. There were several reasons for this quick collapse. At that time the frills of major brand marketing were quite limited, making it more feasible to respond to the self-service gasoline discounters with deep price cuts. Also self-service was restricted to certain areas because of laws and ordinances that were passed to prohibit its growth. Finally, private brand marketing was quite small at the end of World War II and there were relatively few companies by today's standards to spread this method of marketing.

The rebirth of self-service came much more slowly, less spectacularly, and with relatively little publicity. While the momentum gathered very slowly, the base was established for the explosive growth of self-service in the late 1960's and early 1970's. Several of the varying characteristics of self-service for the two time periods are discussed below. One of the big differences was that Pat Griffin and other leaders in the movement followed a policy of regionally diversifying locations. Griffin had a policy of building one, or no more than two, self-service outlets per market, and he scattered these units thinly over the states of Colorado, Arizona, Nevada, Idaho, and Wyoming. Furthermore, his growth could not be called spectacular; he grew from a dozen units at the beginning of the 1960's to about 100 units ten years later. With austere-looking facilities, often in secondary locations, the market for self-service appeared to be rather limited and the competitive reaction was not one of grave concern.

The market circumstances in which self-service grew in the latter 1960's and the early 1970's were also quite different from those of the postwar forties. During the earlier period of self-service there had not been the costly addition of marketing frills which existed by 1970. Over the two intervening decades marketing costs had been layered on and institutionalized by the major brand operations as necessary for doing business. The costs piled on included those for credit cards, trading stamps, brand-name advertising, sales promotions, and elaborate station design. All of these factors drove up the cost of major brand marketing compared to the basic, stripped down operation that Gasamat and others offered. As a consequence, the major brand method of marketing was more vulnerable to a hard-hitting price method of competition by this later time.

A second difference in the rebirth of self service was that gasoline was treated by tie-in operators as just another product line. This greatly reduced the vulnerability of the self-service gasoline tie-in to price wars and to fear of being too price aggressive and catapulting markets into price wars.

A third difference in this period of self-service rebirth was improvements in efficiency of self-service equipment. Equipment advances enabled a central cashier to oversee and control from one to 24 pumps. This was more efficient than the older, Urich type of operation, where women cashiers roller-skated from one car to another collecting money from customers. In addition there were coin and token operations which put the burden of dispensing the gasoline and paying for it largely on the customer.

First Generation Self-Service

Three distinct groups of self-service gasoline marketers made up the first generation. One of these was the new, austere, low-cost, self-serve gasoline outlet. The pioneer in this type of operation was the Gasamat chain. Another of the early low-cost utilitarian outlets was the large-volume Shepherd Brothers' station in Phoenix. Volume was so great at the first unit that a second station was opened adjacent to the first. Shepherd's outlet had three long islands parallel to the road and a central cashier's booth covered by a huge canopy. This layout has become fairly standard for new independent self-serve stations. With little quality self-service competition, this functional but unesthetic unit sold up to half a million gallons of gasoline per month.

Another type of first generation self-service gasoline outlets were those tied in with convenience stores, dairy stores, and other types of retail operations. Very often one island with two pumps would be added to a convenience store operation with the self-service pumps placed as shown in Figure A-2.

One of the problems with tied-in self-service gasoline operations is that often gasoline has been treated as an add-on to an existing business, rather than being developed as an intregal part of the operation. As a consequence, customers frequently find it difficult to obtain the cashier's attention so that the control mechanisms can be set permitting the customer to dispense gasoline. When the dispensing is completed, the customer must walk across part of the parking lot, frequently between parked cars, to enter the convenience store to pay for the gasoline. If the attendant is checking out a grocery customer, the gasoline customer must wait until this transaction is completed before paying. If conditions are good, the dispensing of gasoline from the convenience store can be a reasonably satisfactory experience for the customer. A better physical arrangement is needed for tying self-service gasoline in with the operation of another type of retail outlet. New layouts are being tested by convenience

FIGURE A-2

Alternative Layout for Tie-in Self-Service

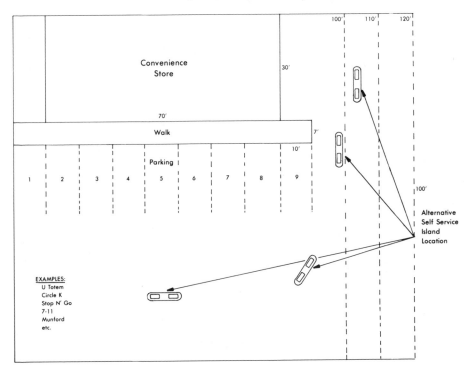

and dairy store operations wanting to move into the self-service gasoline business. One of the newer layouts is shown in Figure A-3. Self-service gasoline is at the very front of the building, near the cashier of the convenience or dairy store. The gasoline customer is now much nearer to the cashier where it is much easier to obtain the store attendant's attention so that the customer can obtain better service. It offers a much better arrangement for the customer to pay for the gasoline. The gasoline customer pays at a window at the front of the store. The store's other customers park on either side of the building and enter the store from doors at the front.

The third type of first-generation self-service operations were converted private brand outlets. Such conversions were handicapped by deficiencies that normally accompany trying to renovate a place designed for one purpose to make it satisfy another set of requirements. Nonetheless, several of the price aggressive private branders made the switch from pumping gasoline and giving other customer services to the self-service method of marketing gasoline. Notable among these was Mary

Self-Service Gasoline Integrated with Other Retail Outlets

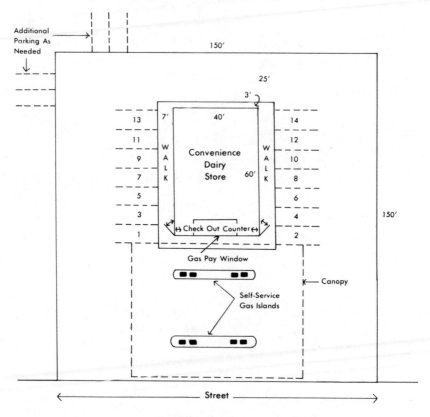

Hudson, who announced plans in July 1971 to convert as many of her 300 stations to self-service as permitted by law.

The three first generation type self-serve operations—new, austere, low cost, self-service units; tie-ins with other stores; and converted private branders—generally had limited marketing appeal because of the physical appearance, choice of location, and layout of the operation. Their customers were often those only interested in the lowest possible price.

Second Generation Self-Service

The second generation of self-service represented an upgrading and commercialization of the self-serve gasoline outlet. These newer units had an improved appearance in both quality and esthetics and had a large canopy covering the pump islands. Some of the leaders in upgrading the quality of independent self-service were Giant located in Phoenix, USA (U Serve Automatic) and Panova based in California, and Autotronics with headquarters in Texas. Others participating in this development included U-Pump and Whiting from Phoenix, Mars from St. Louis, and Raceway based in Alabama.

The layout of a typical second generation self-service gasoline outlet is presented in Figure A-4. The large number of available "gassing points" is one of the features of the second generation self-service. The customer seldom need wait. The cashier's booth is located on the center island, where there are control mechanisms for all the pumps. The operation is covered with a large, well-lighted canopy protecting the self-service customer from the sun or inclement weather.

These operators found that it paid to obtain a better location, to have an attractive design and color scheme, to be well-lighted and to provide customers the protection of a canopy. As the second generation self-service operation caught on, it diminished the effectiveness of the first generation self-service units. For example, Giant's self-serve operation out of Phoenix sold its converted units and opened only huge, imposing self-service stations (see photo, p. 237). As quality competition increased Griffin set about upgrading the appearance and location of his Gasamat facilities. Similarly, Autotronics, the largest of the self-serve operators, changed its emphasis from the low investment, tie-in, self-service operations to the building of attractive, well-located, canopied self-service units (see photo, p. 237).

The second generation operators broadened the appeal of self-service by reaching the economy-quality oriented customer and not simply those interested in the lowest possible price and convenience. They not only reached the customer interested in saving, but also those interested in appearance, convenience, and quality of facilities.

An Analysis of Self-Service in Western Markets

Self-service was pioneered in several of the near western markets by the independent marketers and spread from there throughout most of the rest of the country. Data showing the growth of self-service for Tucson, Phoenix, Salt Lake City, Denver, Reno, and Las Vegas is presented in Table A-1. The data is divided into two categories—the major brand outlets of the large integrated oil companies and the private brands outlets which are predominantly operated by independent marketing companies. As can be observed from Table A-1, by mid-1966, 8 percent of private brand outlets were self-service, while no major brand outlets were self-service. By mid-1969, half of the private brand outlets were selling self-service gasoline in contrast to only 4 percent of major brand stations. By midyear 1971, almost 60 percent of the private brand outlets were dispensing gasoline by self-service in contrast to less than 20 percent of the major brand outlets.[9]

FIGURE A-4

Second Generation Quality Self-Service Layout

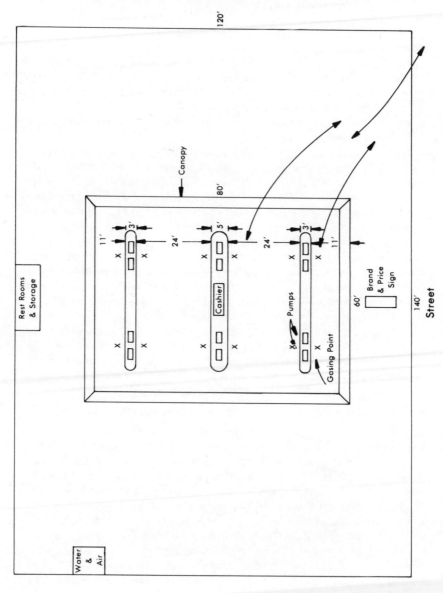

1. Giant's second-generation self-serve station

2. Autotronics' canopied second-generation station

3. American's split-pitched canopy self-serve station with central cashier

4. Shell's experimental ground-up self-serve station

5. Mobil's self-serve plus concession type station

TABLE A-1

Growth of Self-Service in Five Western Markets

City	Type	1966	1967	1968	1969	1970	1971	1972	1973
Tucson	Major Brand	—a	3%	3%	3%	3%	2%	7%	5%
	Private Brand	18%	41	52	62	66	73	77	80
Phoenix	Major Brand	—	—	—	—	2	8	31	23
	Private Brand	—	8	24	58	66	74	83	75
Salt Lake City	Major Brand	—	no data	4	8	14	30	41	32
	Private Brand	10	no data	39	47	66	69	75	70
Denver	Major Brand	no data	no data	2	6	14	20	35	20
	Private Brand	no data	no data	46	54	53	61	64	57
Las Vegas and Reno	Major Brand	—	no data	—	—	1	1	6	3
	Private Brand	4	no data	11	13	25	24	32	44
Average for 5 Cities	Major Brand	—	1	1	4	11	19	30	18
	Private Brand	8	21	39	49	54	59	73	68

a. Indicates that no stations were self-service.

Source: Based upon Lundberg Surveys for the last week of July of the respective year.

The major oil companies did not make their move into self-service until the early 1970's. As can be observed from Table A-1, the relative number of major-brand self-service operations more than doubled from mid-1969 to mid-1970, and almost doubled again from 1970 to 1971. Finally, it increased another 50 percent from 1971 to 1972, to a level where 30 percent of major brand stations had self-serve gasoline. At that time almost three-fourths of the private brand stations sold self-service gasoline.

The difference in the general level of prices for major brand service and self-service, and private brand service and self-service, can be observed from Table A-2. As can be seen from the data, self-service enabled private branders to increase their price discount by approximately two cents per gallon compared to major brand full service. For the five-year period of 1968 to 1972, customers were able to save 5.3¢, 4.4¢, 6.2¢, 4.0¢, and 6.0¢, respectively, in comparison to major brand service price. It was these large customer savings that resulted in the rapid erosion of the major brand market share and the growth of private branders.

TABLE A-2

Major Brand and Private Brand Service and Self-Service
Prices for Regular Grade Gasoline

City	Type	1968	1969	1970	1971	1972
Tucson	Major brand full service	34.9¢	29.0¢	31.6¢	30.5¢	32.5¢
	Major brand self-service	33.9	26.6	28.6	27.4	28.8
	Private brand full service	31.2	26.7	28.4	28.1	29.1
	Private brand self-service	29.6	25.5	27.3	27.0	27.2
Phoenix	Major brand full service	35.0	32.1	33.5	29.1	32.7
	Major brand self-service	32.9	27.9	30.2	26.0	29.4
	Private brand full service	31.2	29.2	28.4	27.3	29.3
	Private brand self-service	29.7	27.7	26.4	25.8	26.7
Salt Lake City	Major brand full service	32.9	34.7	35.2	32.0	36.1
	Major brand self-service	29.6	31.0	31.6	28.6	32.2
	Private brand full service	30.3	31.0	31.9	30.3	32.4
	Private brand self-service	28.2	28.9	29.2	27.2	30.1
Denver	Major brand full service	32.9	29.8	35.8	32.3	34.7
	Major brand self-service	29.6	26.4	32.7	29.3	31.3
	Private brand full service	29.0	27.4	30.9	29.9	29.8
	Private brand self-service	27.0	25.8	28.4	27.9	28.1
Average	Major brand full service	33.9	31.4	34.0	31.0	34.0
	Major brand self-service	31.5	28.0	30.8	30.4	30.4
	Private brand full service	30.4	28.6	29.9	28.9	30.2
	Private brand self-service	28.6	27.0	27.8	27.0	28.0

Source: Based upon Lundberg Surveys for the last week of July of the respective year.

Several of the companies that played a major role in the development of self-service are listed in Table A-3.

Majors' Response to Self-Service

The prevalent attitude of the majors seems to be regret that self-service ever occurred and the wish that it would go away. What self-service represented to the major oil companies was a devastating method of competition to their branded, nonprice method of marketing. The more they were drawn into competition with self-service, the more problematic became their own costly, overbuilt marketing structure.

Because of their heavy investment in the traditional market, the major oil companies were slow to meet the challenge of self-service. Their response was primarily defensive and consisted of converting an outer or side island to self-service, while continuing to stress service at the others.

APPENDIX

TABLE A-3

Self-Service Hall of Fame

Category	Self-Service	Tie-In Self-Service	Converters to Self-Service
Early self-service: (1947–51)			
Pioneer	Urich		
Early adopter	Powerine		Golden Eagle
Rebirth self-service: (1960–72)			
Pioneer	Gasamat	Timms	
Early adopter	Shepherd	Nu Way	Hudson
		Tenneco	
Commercializer	Autotronics	Autotronics	Thrifty
	U.S.A.	Pioneer	World
	Giant	Romaco	
	Powerine	Egerton	
	Kickapoo	Marshall	
	Christopher	Circle K	
	Raceway	Munford	

The majors' emphasis on split island self-service in comparison to the independents' stress on total service can be observed from Table A-4. By 1972 more than 85 percent of the major brand self-serve stations offered split service. In contrast, about 35 percent of the private brand self-service outlets were split island operations. Indicative of the majors' defensive use of self-service was their rapid withdrawal from self-service from 1972 to 1973 as the gasoline shortages developed. From midyear 1972 to 1973, major brand self-service fell from 30 percent to 18 percent of the major brand outlets (see Table A-4). In comparison, private brand self-service fell little from 73 percent to 68 percent of the stations surveyed.

One of the problems of the split-service type operation has been that it is difficult to do a good job with both service and self-service in the same station because of conflicting characteristics. As a consequence one type of customer is not treated as well as he would be in an "unmixed" operation. In some cases both service and self-service customers may experience a diminution in the level of service being offered on a split-island basis.

The majors' response to self-service was fundamentally not to panic. As the pressures of self-service mounted, several of the major oil companies started limited marketing tests. American Oil Company was one of the first to do this and, in early 1967, it converted seven of its stations in Tucson to self-service (see Table A-4). What followed was a dogfight among the majors over price differential for service and self-service

TABLE A-4
Split and Full Self-Service of Major and Private Brand Outlets

City		1966 SSS	1966 SS	1967 SSS	1967 SS	1968 SSS	1968 SS	1969 SSS	1969 SS	1970 SSS	1970 SS	1971 SSS	1971 SS	1972 SSS	1972 SS	1973 SSS	1973 SS
Tucson	Private brand	0%	18%	8%	33%	16%	36%	15%	47%	0%[a]	66%	0%[a]	73%	13%	64%	11%	69%
	Major brand	0	0	0	3	0	3	1	2	0[a]	3	0[a]	2	3	4	2	3
Phoenix	Private brand	0	0	3	5	9	15	18	40	0[a]	66	0[a]	74	12	71	4	71
	Major brand	0	0	0	0	0	0	0	0	0[a]	2	0[a]	8	25	6	18	6
Salt Lake City	Private brand	7	4	no data	no data	24	14	29	18	41	25	39	30	34	41	29	42
	Major brand	0	0	no data	no data	3	0	7	1	12	2	26	4	34	7	23	8
Denver	Private brand	no data	no data	no data	no data	20	26	30	24	28	25	38	23	47	17	36	19
	Major brand	no data	no data	no data	no data	2	0	0	0	14	0	20	0	35	0	14	1
Las Vegas & Reno	Private brand	0	4	no data	no data	0	15	0	13	2	19	0	27	3	29	2	42
	Major brand	0	0	no data	no data	0	0	0	0	0	0	1	0	3	1	3	0
Average	Private brand	3	5	5	16	18	21	24	26	30[b]	24[b]	33[a]	26	26	47	18	50
	Major brand	0	0	0	1	1	0	3	1	11[b]	0[b]	18[a]	1	26	4	14	4

a. No distinction between self-service and split, self-service, stations.
b. These figures are compiled without distinction in data for Phoenix and Tucson.
Source: Lundberg Survey Data.

operations. Some major brand competitors were unwilling to give American self-service any price edge and Tucson witnessed almost continuous price turmoil from 1967 to 1972.

The violent reaction to American's experiment was certainly a deterrent to other majors beginning tests of self-service. There were no other particularly noteworthy tests of self-service by major oil companies until early 1969, when Humble announced a widespread self-service test of 25 stations. Along with Humble, other major oil companies started extensive experiments of their own and the majors entered the self-service game on a split-island basis.

Moving from earlier emphasis on conversion of existing units, Humble in 1971 and 1972 experimented with some newly constructed "ground, up" self-service stations and the major brand second generation self-service experiments were beginning to expand. Humble's approach was to play down decor and to have functional appearing, multiple island, canopied self-service gasoline outlets tied in with a carwash operation. American opened several large, attractive, split-pitched canopy, three island, central cashier type self-service stations in Salt Lake City (see photo, p. 237). Shell proceeded to open several experiments with spacious and attractive self-service units in the western part of the United States (see photo, p. 237). The experiments' success led Shell to make a major commitment to future construction of self-service units. Mobil took a somewhat different tack with their new units. Instead of the gasoline-only approach of Shell and American, Mobil used a central cashier located in a long, narrow, concession unit where customers could also buy a limited line of grocery items as they paid for their gasoline (see photo, p. 237).

The second generation major brand self-service development was just getting under way when the gasoline shortage hit. Had it not been for the shortage, the second generation major brand self-service units would have been rapidly expanded.

Violent Price Wars Accompany Self-Service

The penetration of self-service in new marketing areas was very often accompanied by severe price wars. Traditional service operations were unwilling to give up substantial volume to the low-price self-serve operators, and when volume losses grew they reduced their prices. As a consequence, markets where self-service penetration was growing were often those with chronic price wars. Tucson was one of the earliest markets where self-service made substantial inroads and, as noted, it experienced violent price warring from 1966 through mid-August 1972.

The impact that the growth of self-service had on prices in Phoenix can be observed from an analysis of the weekly changes in prices for the majors and independent private brands as presented in Figure A-5. During 1966 and early 1967, there were only limited experiments with self-service in Phoenix and, as can be observed from the data, there was relative price stability. From the limited experiments at the beginning of this period, the independents jumped into self-service, so that, by mid-1969, close to 60 percent of the independents were offering self-serve gasoline. The chart similarly shows the growing instability of prices from 1966 through 1969. Prices were very unstable during 1969 and there were eight major price war periods in contrast with four during the previous year, three during 1967, and two during 1966.

FIGURE A-5

*Fluctuation on Major Brand and Private Brand Prices
for Regular Gasoline in Phoenix from 1966 to 1972*

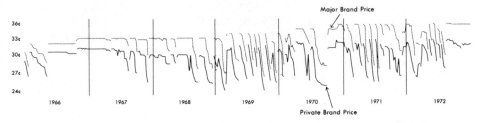

Source: Developed from weekly Lundberg Price Survey.

The Phoenix market seemed to adjust to the rapid movement of independents into self-serve during 1969 and prices stabilized somewhat during 1970, when there were only four major breaks in prices, in contrast to eight in 1969. During 1971 and 1972 the major oil companies made their big push into self-serve so that, by mid-1972, one-third of the major brand stations were dispensing gasoline by the self-serve method, as contrasted with only 2 percent by mid-1970. With the majors' move into self-serve, prices started once again to gyrate. During 1971 there were eight major breaks in the normal price levels and "normal prices" were never maintained for more than two to three weeks before prices catapulted downward by six to eight cents per gallon. Prices continued their yo-yo pattern in 1972 and there were five major breaks in prices over the first seven months of the year.

The bitter reception of self-serve in Phoenix was similar to what happened in other markets throughout the United States where self-service was making significant market inroads. Price warfare was unable to stop the growth of self-service because of its favorable economics and other attractive features. Self-service grew in Phoenix until more than 80 percent of the private brand marketing outlets had self service by mid-1972 and approximately one-third of the major brand stations were offering some form of self-serve gasoline to their customers (see Table A-1). As earlier noted, estimates placed self-service as high as 60 percent of the gasoline sold in Phoenix in 1972.

Third Generation Self-Service

It is shortsighted to believe that the gasoline revolution is going to stop with second generation self-service. Competition permitting, the future will see a great deal of high volume, quality second generation self-service along with self-service tie-in operations. The challenge of the future to the innovative marketer will be a third generation self-service. The third generation will find the high volume, high impact, canopied, second generation self-service units incorporated into planned retail development.

The innovative opportunity of the future will be to take the 150,000, 200,000, and 300,000 gallon per month self service units and integrate them into planned retail development for the mobile shopper. The objective will be to develop the center around the mobile shopper's need for making quick purchases. Gasoline often is, and in the future may be increasingly, purchased as part of a more comprehensive shopping experience. Many gasoline customers purchasing gasoline stop at other retail outlets for "pick-up" items before or after buying gasoline. The objective is to design the quick purchase center for the mobile shopper so that the operations complement and reinforce each others' drawing power.

Self-service gasoline will be the central activity in the center for the mobile shopper because of its importance and nature of the traffic pattern flow. The large volume, canopied, self-service gasoline operations will be flanked to the side and back by a variety of small, convenience type retail outlets. One possible example of this new type of integrated merchandising of gasoline with other products purchased by the mobile shopper is presented in Figure A-6. The canopied, high volume, self-service gasoline operation is in the middle and the three islands are laid out parallel to the major artery. On one side is a convenience and liquor store and on the other side a laundry pick-up, chicken carry-out, and doughnut store. To the back additional overflow and employee parking is provided, and, if the area warrants, additional retail operations requiring less exposure can be developed. These might include a tire store, pharmacy, hardware store, and possibly service establishments such as a barber shop, beauty parlor, or real estate office.

The response of many second generation self-serve operators to the third generation concept is predictably going to be quite negative. As some have already suggested, they have enough challenge simply trying to master self-service gasoline, so why add the headaches of trying to plan a more complex type of operation? The reason this is important for those who intend to stay at the forefront of the self-service evolution is that specifications for a good, second generation self-service operation are easy to write. But, as more and more of these second generation units are opened, the competitive advantage and good rates of return will quickly erode. This, however, will not be true for the more comprehensive, third generation self-service concept which takes more skills to develop and is

FIGURE A-6

Third Generation Self-Service
Gasoline in Planned Convenience Center

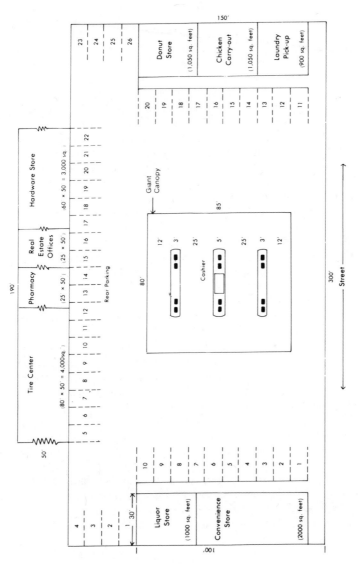

not easily imitated. Those that master this approach will have important competitive advantages over many straight self-service operations, including more customer drawing power and fuller utilization of real estate properties.

There have already been some limited experiments of integrated gasoline marketing and more complex operations are being discussed and planned. Giant in Phoenix has recognized the loss of opportunity for profits and competitive advantage as their customers stop at other quick purchase outlets before or after buying gasoline. New Giant stations are being constructed with Maxi-Marts which provide their gasoline customers an opportunity to purchase other products while on the premises of the gasoline operation. Giant's Maxi-Mart brings together under one roof a large convenience store, a film and camera center, and a laundry pickup operation. Tire stores have been recently added to the mix. Further plans are being made to try to incorporate prepared foods into the planned real estate development.

Tenneco has also been successful in mastering the integrated approach to merchandising of self-service gasoline and convenience grocery outlets as separate and distinct businesses on the same property. The rate of return from their integrated marketing outlets has been exceptional in comparison to that earned on major brand marketing. One of the important reasons for this is that Tenneco has two profitable businesses on the same property that have permitted them to obtain good locations and to more fully utilize the locations than their competitors. Tenneco has also found that their combined operations enjoy the synergistic effect of each operation helping to increase the drawing power of the other.

The development of third generation self-service in planned retail development is primarily limited by the imagination of management, the size of available properties, and the nature of the surrounding market. In some units, Tenneco has gone further and added laundry and other types of retail outlets to their gasoline and convenience store combinations.

There are smaller and less conspicuous operators who have successfully experimented with combination operations and who are planning ways for integrating high volume self-serve gasoline with other complimentary retail outlets into planned real estate developments.

This third generation approach to self service because of its economical customer benefits and its sophistication could very well bring a maturing to the marketing of gasoline which is based upon sound fundamentals of merchandising rather than trivial manipulation of a company's image through appearance, advertising, and promotion programs. As mass merchandising belatedly comes of age in gasoline marketing, more gasoline volume will be efficiently funnelled through fewer and fewer high volume, integrative, super self-service units. The outmoded, low volume, conveniently located, expensive, high priced branded service stations will suffer as a consequence and could well go the way of the corner grocery store.

Summary and Conclusion

The rapid growth of self-service and its revolutionary impact on traditional service outlets came to a screeching halt in the early part of 1973 with the beginning of widespread gasoline shortages. This represented the second time since World War II that self-service had been launched and strenuous opposition was mounted against it.

Second and third generation self-service operations, plus convenience store tie-ins of the first generation, will definitely capture a much larger share of the market if and when competition returns to the marketing of gasoline. The public will be the beneficiary if the independents are once again able to purchase a competitively priced gasoline, for the independent, private brand marketers can operate for as little as one-half of what it costs the majors to market gasoline. The savings to the public of making competitively priced gasoline available to the independent private brand marketers amounts to several billion dollars per year.

The growth of self-service in the later 1960's and early 1970's differed markedly from that occurring twenty years earlier, at the end of the 1940's and early 1950's. The rebirth of self-service has been more widespread with deeper penetration than that occurring at the end of World War II. The greater market entrenchment of self-service today should permit it to hold on for a considerable period of time until a competitive supply market for gasoline returns. If it were not for a high degree of vertical integration and control over supply of gasoline by the giant petroleum companies, supply of gasoline would have continued to have been diverted from the very inefficient major brand method of marketing to the supercompetitive self-service approach led by the independent private brand marketers.

There are glowing forecasts for self-service with estimates of up to 50 percent of the gasoline being sold through self-service outlets by 1980. Whether or not this potential is ever achieved, and the public benefits from the efficiencies of self-serve, will depend upon the return of competition to the marketing of gasoline. The innovators and leaders of the self-service gasoline revolution must not be cut off from gasoline supplies or priced out of the market by being forced to pay unfair prices, if the public is going to enjoy the benefits from the potential self-service gasoline revolution.

Notes

CHAPTER 1

1. U.S. Congress, Committee on Interior and Insular Affairs, *The Gasoline Shortage: A National Perspective*. A Background Paper prepared by the Congressional Reference Service, Committee Print, 93rd Congress, 1st Session (June 19, 1973), p. 35.

CHAPTER 2

1. Warren C. Platt, "40 Great Years—The Story of Oil's Competition," *National Petroleum News* (March 9, 1959), p. 42.

2. John G. McLean and Robert William Haigh, *The Growth of the Integrated Oil Companies* (Boston: Graduate School of Business Administration, Harvard University, 1954), p. 270.

3. *National Petroleum News Factbook*, Mid-May 1973, p. 60.

4. Jeremy Main, "Meanwhile Back at the Gas Pump—A Battle for Markets," *Fortune* (June 1969), pp. 108–109.

CHAPTER 3

1. Fred C. Allvine and James M. Patterson, *Competition Ltd.: The Marketing of Gasoline* (Indiana University Press, 1972), p. 109.

2. Ibid., p. 110.

3. Ibid., p. 113.

4. Ibid., p. 128.

5. Ibid., p. 129.

6. *National Petroleum News Factbook*, Mid-May 1970, p. 32.

CHAPTER 4

1. Allvine and Patterson, *Competition Ltd.*, pp. 217–224.

2. Ibid., p. 245.

3. "Large Firms Improve Crude Oil Base," *Oil & Gas Journal* (January 10, 1972), p. 30.

4. "Socal's Particular About Growth Area," *Oil & Gas Journal* (November 1, 1965), p. 28.

5. "Cost Rise Trims Eastern Hemisphere Profits," *Oil & Gas Journal* (July 9, 1973), p. 45.

6. Ronald Koven and David B. Ottaway, "Oil Companies to Become Maid-in-Waiting?" Atlanta *Journal and Constitution* (December 16, 1973), p. 6E.

7. "Gulf Tries Merger to Save its Profit Problems," *Business Week* (November 13, 1973), p. 10.

8. Statement of the American Petroleum Institute for Hearings before the Committee on Interior and Insular Affairs, U.S. Senate, October 5, 1973, Answer to Question 42f, p. 3.

9. "Drilling Costs Hit High of $19.03/ft.," *Oil & Gas Journal* (December 25, 1972), pp. 46–47.

10. "Higher Off-Shore Bonuses Spell Tougher Oil Economics," *Oil & Gas Journal* (January 1, 1973), p. 6.

11. "Chase Sees $1 Trillion Oil Spending Need," *Oil & Gas Journal* (March 26, 1973), p. 40.

12. "Bank Sees Turning Point for Oil Capital," *Oil & Gas Journal* (November 5, 1973), p. 36.

13. "OPEC to the Oil Industry: 'You've Got Yourself a Partner,'" *Forbes* (March 15, 1972), p. 28.

14. "Gulf Tries Merger," p. 110.

15. "Higher Offshore Bonuses," p. 6.

16. Frank Breese, "Behind Our Headlines," *National Petroleum News* (January, 1973), p. 5.

17. Keith Fanshier, "The Look Downstream," *The Oil Daily* (January 3, 1972), p. 4.

18. "Gulf Tries Merger," p. 114.

19. Ibid.

20. "Rack Pricing Plan Inevitable, says Burnap," *U.S. Oil Week & Car Wash Week* (February 12, 1973), p. 1.

21. Speech by Robert T. McCowan to the Colorado Oil Marketers Association (October 4, 1971), pp. 5, 6, and 8.

22. "Arco Offers Southern States to Jobbers," *U.S. Oil Week & Car Wash Week* (November 22, 1971), p. 2.

23. "If a Major Can't Make It in a Market," *National Petroleum News* (February 1972), p. 43.

24. "Arco President Outlines How Pruning Will Better Company Profits," *National Petroleum News* (January 1972), p. 24.

25. "Quote of the Week," *U.S. Oil Week & Car Wash Week* (June 12, 1972), p. 1.

26. "Why Phillips Quit the Northeast," *National Petroleum News* (August 1972), p. 85.

27. *Gulf Oil Corporation, 1972 Annual Report*, p. 12.

28. "Rack Pricing Plan Inevitable," p. 1.

29. "Sun Pulls Out of 8 States, Cites Unprofitability," *National Petroleum News* (March 1973), p. 10.

30. "Top Conoco Marketer Proposes New Pricing System for Gasoline," *The Oil Daily* (June 26, 1971), p. 2.

31. "Gasoline Prices Plummet; Conoco, Phillips Drop 'No Support' Tests," *National Petroleum News* (April 1971), p. 20.

32. Bill Mullins, "Conoco Marketer Advises Iowa Jobbers: Price Props to be Around awhile," *The Oil Daily* (February 18, 1971), pp. 1, 4.

33. "New Humble's No. 1 Marketer Looks at Oil Marketing Today," *National Petroleum News* (February 1960), p. 92.

34. "They're Holding Feet to the Fire at Jersey Standard," *Fortune* (July 1970), p. 130.

35. "A Marketer Looks at Challenge," *The Oil Daily* (November 12, 1970).

36. *National Petroleum News* (April 1974), p. 11.

CHAPTER 5

1. "Gulf Tries Merger," p. 110.

2. "Oil: The Created Crisis," reprints, *Philadelphia Inquirer,* from a three-part series commencing May 22, 1973.

CHAPTER 6

1. Memoranda of the Office of Oil and Gas dated January 12 and 20, 1972, questioned the intention of Louisiana to withhold some production and not go to maximum production on national defense grounds. There was fear that Texas and other states might decide on the same course of action. An investigation of this policy was recommended by staff of the Office of Oil and Gas. No determination was made concerning whether an investigation was actually undertaken.

2. Robert Plett, *Memorandum for the Record for the Office of Emergency Preparedness,* on "Trip Report—Texas Railroad Commission," March 21, 1972, p. 3.

3. Ibid.

4. Elmer F. Bennett, Assistant Director of OEP, *Memorandum for the Director,* March 23, 1972.

5. Memorandum by General George A. Lincoln to Peter Flanigan, dated March 21, 1972.

6. Memorandum prepared by William Truppner for the Director, dated March 23, 1972.

7. Minutes of the Oil Policy Committee prepared by Assistant Director of Oil Imports in the Office of Oil and Gas, dated April 25, 1972.

8. *Memorandum for the Record of the OEP* prepared by Philip L. Essley, Jr., dated July 28, 1972.

9. H. H. Roberts, Jr., *Memorandum for the Record of the OEP,* dated August 16, 1972.

10. Phillip L. Essley, Jr., *Memorandum for the Record,* Office of Emergency Preparedness, August 9, 1972, concerning a meeting with Citgo representatives on August 3, 1972.

11. Robert Plett, *Memorandum for the Record for the OEP,* dated April 27, 1972.

12. General George A. Lincoln, *Letter to New England Governors' Conference,* dated April 26, 1972.

13. *Memorandum for the Record* to General Lincoln, dated May 3, 1972.

14. Robert E. Plett, *Memorandum of Meeting with Humble Oil Company to Assess the Fuel Oil Supply Situation,* dated July 19, 1972.

15. Robert Shepherd, *Memorandum* prepared for General Lincoln, dated July 31, 1972.

CHAPTER 7

1. *Hearings on Adequacy of Home Heating Oil Supply,* Subcommittee on Small Business of the Committee on Banking, Housing, and Urban Affairs, U.S. Senate, 92nd Congress, 2nd Session, September 19 and 20, 1972, p. 14.

2. Ibid.

3. Ibid., p. 20.

4. Ibid., p. 13.

5. Ibid., p. 14.

6. Ibid., p. 106.

7. Ibid., p. 102.

8. *Hearings on the Inadequacy of Petroleum Supplies and its Repercussion on Small Businesses,* Subcommittee on Small Business Problems, Select Committee on Small Business, House of Representatives, 92nd Congress, 2nd Session, September 27 and 28, 1972, p. 72.

9. Ibid., p. 87.

10. Ibid., p. 94.

11. Ibid., p. 54.

12. Ibid., p. 58.

13. Ibid., p. 59.

14. Howard Roberts, Memorandum prepared of meeting between OEP and Humble Oil Executives on November 3, 1972, dated November 8, 1972.

15. Howard Roberts, Memorandum of meeting with Texaco on November 8, 1972, dated November 14, 1972.

16. Howard Roberts, Memorandum of meeting with Sun Oil Company on November 9, 1972, dated November 14, 1972.

17. Howard Roberts, Memorandum of a meeting with Shell Oil Company on November 14, 1973, dated November 15, 1973.

18. General George Lincoln, *Memorandum for the Record,* November 29, 1972.

19. Ibid.

20. Howard Roberts, *Memorandum for the Record* of a meeting with Shell Oil Company on February 1 dated February 13, 1973.

21. Robert E. Shepherd, *OEP Memorandum* to Darrell Trent, dated January 24, 1973, which deals with the proceedings of the NOJC meeting of January 23, 1973.

22. *Oil & Gas Journal* (March 22, 1971).

23. "Refinery Capacity Now Number One Issue in Washington," *Oil & Gas Journal* (November 15, 1971), p. 90.

24. Leo R. Aalund, "Refinery Trends are Submerged by Flood of Uncertainty," *Oil & Gas Journal* (March 25, 1972), pp. 108–125.

25. Leo R. Aalund, "Work Start Needed Soon to Avoid '75 Gasoline Gap," *Oil & Gas Journal* (March 25, 1972), pp. 29–31.

CHAPTER 8

1. "Staff Study of the Oversight and Efficiency of Executive Agencies with Respect to the Petroleum Industry, Especially as it Relates to Recent Fuel Shortages," Permanent Subcommittee on Investigations, United States Senate, November 8, 1973, pp. 70, 89, 93.

"Fuel Shortages," Committee on Interior and Insular Affairs, United States Senate, February 22, 1973, Part 2, pp. 561, 632, 664, 668, 671, 699–701.

"Petroleum Product Shortages," Committee on Banking, Housing and Urban Affairs, United States Senate, May 7–11, 1973, pp. 233, 252, 261–63.

2. "Fuel Shortages," pp. 667, 673, 700. "Petroleum Product Shortages," pp. 228, 230, 240, 263, 269, 289. The Atlanta *Constitution,* Thursday, January 10, 1974, p. 22A.

3. The majors responding to the questionnaire were Mobil, Amoco, Shell, Phillips, Sun, Union Oil of California, Standard Oil of Ohio, Cities Service, Getty, and Marathon.

4. "Train: More Flexibility, Time Needed," *The Oil Daily* (October 10, 1973), p. 1.

5. "Staff Study of . . . Recent Fuel Shortages," p. 70. "Petroleum Product Shortages," pp. 229, 289.

6. Donald R. Bartlett and James B. Steele, "Oil Firms Sell Abroad, U.S. Pays," *Philadelphia Inquirer,* Special Reprint, July 22, 1973, pp. 2–3.

7. "North Carolina's Self Serve King Shouts, 'Olé'," *National Petroleum News* (June 1970), p. 64.

8. "Critics Cold to Coin-Op Struggles as Battle in North Carolina Heats," *National Petroleum News* (September 1972), p. 32.

9. *Platt's Oilgram Price Service,* Vol. 52, No. 16 (January 23, 1974), p. 3.

10. *Touhey Report,* January 14, 1974, p. 3.

11. *NPN Bulletin,* January 21, 1974, p. 2.

12. *Platt's Oilgram Price Service,* Vol. 52, No. 13, p. 1.

13. Ibid.

14. *U.S. Oil Week,* January 14, 1973, pp. 2, 6.

15. Advertising of Shell Oil Company explaining reason for higher 1973 profits, The Atlanta *Constitution* (January 28, 1974), p. 3A.

16. "Marketing Rumors Fly on Price Controls," *The Oil Daily* (August 14, 1972), p. 1.

17. "Small Marketers Ask Supply Allocation," *The Oil & Gas Journal* (May 7, 1973), p. 3.

18. "Oil Firms Reported Willing to Ration," The Atlanta *Journal* (June 7, 1973), p. 17A.

19. "Mandatory Allocation Looms Inevitable," *The Oil & Gas Journal* (June 18, 1973), p. 47.

20. "Are Allocations Working? Well, Nobody Really Wants to Say So," *National Petroleum News* (July 1973), p. 10.

21. "Oil Shortage and Mandatory Allocations," House Committee on Interstate and Foreign Commerce, July 12, 1973.

CHAPTER 9

1. U.S. Congress, Committee on Government Operations, Permanent Subcommittee on Investigations, *Investigation of the Petroleum Industry,* A Preliminary Federal Trade Commission Staff Report on Its Investigation of the Petroleum Industry, Committee Print, 93rd Congress, 1st Session (July 1973), p. 6.

2. Ibid., p. 13.

3. M. A. Adelman, *The World Petroleum Market* (Baltimore: The Johns Hopkins Press, 1972), p. 81.

4. There is a well developed literature here. The technically interested reader should see: F. Machlup and M. Tabor, "Bilateral Monopoly, Successive Monopoly, and Vertical Integration," *Economica,* Vol. XXVII (May 1960); R. Bork, "Vertical Integration and the Sherman Act: The History of an Economic Misconception," *University of Chicago Law Review,* Vol. XXII (Autumn, 1954); J. J. Spengler, "Vertical Integration and Antitrust Policy," *Journal of Political Economy,* Vol. LVIII (August 1950); M. Adelman, "Integration and Antitrust Policy," Harvard Law Review, Vol. LXIII (1949); G. E. Hale, "Vertical Integration," *Columbia Law Review,* Vol. LXIX (1949); F. M. Scherer, *Industrial Market Structure and Economic Performance* (Chicago: Rand McNally, 1970), pp. 69–71.

5. *U.S.* v. *United States Steel Corp. et al.,* 251 U.S. 417 (1920).

6. For a list of the constituent and successor companies and their spheres of operation, see George S. Gibb and Evelyn H. Knowlton, *The Resurgent Years, 1911–1927* (New York: Harper and Bros., 1965), pp. 8–9.

7. Scherer, p. 467.

8. Ibid.

APPENDIX

1. *National Petroleum News,* June 1971, p. 62.
2. Alan Grott, *PMEF Super Seminar #1,* December 1, 1972, p. 3.
3. *National Petroleum News,* June 1971, p. 62.
4. *National Petroleum News,* March 1972, p. 82.
5. *Oil Week,* September 18, 1972, p. 8.
6. *National Petroleum News,* mid–May 1973, p. 98.
7. Peter Britten, *PMEF Super Seminar #1,* December 1, 1972, p. 3.

8. For a more complete discussion of early self-service see Allvine and Patterson, *Competition Ltd.,* p. 76.

9. Less than 20 percent of the private brand outlets represent conversions of major brands such as Phillips and Continental to private brands including Blue Goose, Jiffy, Econo, and Nugget.

Index

Abourezk, Sen., 152
Addabbo, Rep. Joseph P., 139, 199
advertising, 3, 26, 44
Alaska, 60
Alaskan oil pipeline, 6, 8, 60
Amerada-Hess, 168, 203
American Oil Co., 68, 70, 168, 240
American Petrofina, 21, 70, 185, 203
American Petroleum Institute, 59, 62, 133
American Petroleum Refiners Association, 121
Arab Oil Embargo, 53
Arco, 168, 172, 178, marketing strategy, 69–70, refinery capacity utilization, 94, 95
Armour Oil Co., 172, 174
Ashland Oil Co., 66, 121, 132, 203, appeals to government, 151, 165
Atlantic-Richfield, 60, 142. *See also* Arco
Autotronics, 21, 171–73, 186, 235, 240

"balancing wheel" role, Texas and Louisiana, 110–11, 124, U.S. government, 209
Big Eight, 3–4, 18n
Big Fifteen, 18n
bonus bidding, 60
brand image, 26, 46
brand preference, 26
British Petroleum Co. (BP-Sohio), 51, 54, 68, 70, 178

Card, Annon, 153
cartelization of crude production, 28–29
car wash promotions, 44
Catmull, Kenneth, 200

Certified, 21
Chevron, 115
Cities Service (Citgo), 105, 168, earnings, 64, 203, and OEP, 126–27, 140–41, 157
Clark Oil Co., 115, 119–20, 124, 151
Coastal States Oil, 118, 121, 124, 153, 185
competition: intertype. *See* intertype competition; intratype. *See* intratype competition; non-price, 25–27, 44
Congress. *See* U.S. Congress
Continental Oil Co. (Conoco), 64, 66, 68, 72, 128, 168
Cost of Living Council, 165
costs, exploration and development, 49
credit cards, 44
Crown Central, 177, 178, 182–85
crude oil:
  days' supply, 98–99, 100–101, feed stocks, 98–102, 149, 153, imports, 100, 102, 106, 109, 117–18, 119, 130, 133–34, 149–50, 158, Oil Import Program, 15, operations, 47–48, 207, OPEC effect on profits, 56, price levels, 48, 52–53, 58, prices, freeze on, 49, 59, production, 28–30, 57–61, 111, 213, 224, "profit nerve center," 79, 216, profits subsidize downstream operations, 62–67, 76, runs to stills, 83, sales reduced, 210, shortages, 100, 113, 115–21, 124–27, 167, 185, supplies, 98, 100–101, 132–33, 148, 150–51, 209

dealers, 40. *See also* independent marketers, majors, price marketers
debottlenecking, 166, 166n
demand prorationing, 111–12
Dergy, 21